A COMPENDIOUS MANUAL

OF

QUALITATIVE CHEMICAL ANALYSIS.

BY

CHARLES W. ELIOT,
PROFESSOR OF ANALYTICAL CHEMISTRY AND METALLURGY,

FRANK H STORER,
PROFESSOR OF GENERAL AND INDUSTRIAL CHEMISTRY,

IN THE MASSACHUSETTS INSTITUTE OF TECHNOLOGY.

NEW YORK:
D. VAN NOSTRAND, PUBLISHER,
23 MURRAY STREET AND 27 WARREN STREET.

1869.

PREFACE.

The authors have endeavored to include in this short treatise enough of the theory and practice of qualitative analysis, "in the wet way," to bring out all the reasoning involved in the subject, and to give the student a firm hold upon the general principles and methods of the art. It has been their aim to give only so much of mechanical detail as is essential to an exact comprehension of the methods, and to success in the actual experiments. Hence, the multiplication of different tests or processes, having essentially the same object, has been purposely avoided. For the same reason, none of the rare elements are alluded to. The manual is intended to meet the wants of the general student, to whom the study is chiefly valuable as a means of mental discipline, and as a compact example of the scientific method of arriving at truth. To professional students who wish to make themselves expert analysts, this little book offers a logical introduction to the subject—an outline which is trustworthy as far as it goes, but which needs to be filled in and enlarged by the subsequent use of some more elaborate treatise as a book of reference. Prof. Johnson, of Yale, has supplied this need with his excellent edition of Fresenius's comprehensive manual.

The authors believe that they have put into the following pages as much of inorganic qualitative analysis as is useful for training, and also as much as the engineer, physician, agriculturist, or liberally educated man needs to know. The book has been written for the use of classes in the Institute of Technology, who have already studied the authors' "Manual of Inorganic Chemistry." It is simply an implement devised to facilitate the giving of thorough instruction to large classes in the laboratory. It is the authors' practice to examine their classes orally every four or five exercises, in order to secure close attention to the reasoning of the subject. With this exception, the art is studied exclusively in the laboratory, tools in hand. Fifty laboratory exercises of two hours each have proved sufficient to give their classes a mastery of the subject as it is presented in this manual.

It is scarcely necessary to say that this little work is a compilation from well-known authorities, among which may be particularly mentioned the works of Galloway, Will, Fresenius, and Northcote & Church.

BOSTON, *April*, 1869.

TABLE OF CONTENTS.

PART II.

PART III.

QUALITATIVE ANALYSIS.

INTRODUCTION.

1. Qualitative Analysis, in the widest sense of the term, is the art of finding out the elements contained in compound substances. This general definition has important limitations in practice. In the first place, the art, as commonly taught, applies almost exclusively to mineral, or inorganic, substances, and touches only incidentally upon the multifarious compounds of carbon with hydrogen, oxygen, nitrogen, and a few other elements, which form the subject-matter of that branch of chemical science called organic chemistry. Again, the analysis of gases constitutes a distinct branch of analysis, requiring methods and apparatus of its own, and therefore to be most advantageously studied by itself. These deductions made, there remains the analysis of inorganic solids and liquids, which is in fact the main subject of qualitative analysis in the present technical sense of the term.

Of the sixty-five recognized chemical elements, only the thirty-four most important are embraced in the systematic course of this manual. Means of detecting a few other less common elements are incidentally given ; but the elements which are so rare as to be at present of little interest except to the professional chemist or mineralogist, are not alluded to.

2. Some previous knowledge of general chemistry is essential to the successful study of qualitative analysis. It is assumed that the student knows something of the common elements and of their most important combinations, that he is familiar with the principal laws which govern chemical changes, and that he possesses a certain skill in the simplest manipulations. The tools and operations employed in qualitative analysis are few and simple; but neatness, method in working, and a vigilant attention even to the minutest details, are absolutely essential. As the various substances used or produced in the operations of analysis will not be particularly described, the careful student will keep at hand some text-book on general chemistry, to which he can constantly refer to refresh his recollection of the formulæ and physical and chemical properties of the substances referred to. It should be observed that it is often very difficult—in fact, impossible in the present state of knowledge—to express in exact equations the involved or obscure reactions which occur in complex mixtures during the operations of analysis. It is a useful exercise for students to write out in equations the simpler chemical changes which occur in analysis; but when the attempt is made to put a complex reaction into numerical symbols, the equations are apt to express either more than we know, or less.

3. Although the detection of the elements contained in compound substances is the ultimate object of analysis, it is only by exception that the elements themselves are isolated, and recognized in their uncombined condition. An element is generally recognized through some familiar

compound, whose apparition proves the presence of all the elements it contains, just as the presence of any word upon this page makes it sure that the letters with which it is spelt are imprinted there. If, as the result of a definite series of operations upon some unknown body, the hydrated oxide of iron be produced, no iron having been added during any stage of the process, the proof of the presence of iron in the original body is quite as certain as if the gray metal itself had been extracted from it. If some well-known sulphate, like sulphate of lead, or of barium, for example, result from a series of experiments upon some unknown mineral, it is certain that the mineral contained sulphur; provided only that no sulphur has been introduced in any of the chemical agents to whose action the mineral has been submitted.

The compounds through which the elements are recognized are necessarily bodies of known appearance, deportment, and properties. They are, in fact, bodies of various, though always definite, composition; oxides, sulphides, chlorides, sulphates, and many other salts, are thus made the means of identifying one or more of the elements which they contain. The object of the analyst is to bring out, from the unknown substance, by expeditious processes and under conditions which admit of no doubt as to their testimony, these identifying compounds, with whose appearance and qualities he has previously made himself acquainted. As he follows the course of experiments laid down in this manual, the student will gradually acquire, with the aid of frequent references to a text-book of general chemistry, that stock of information concerning the identifying compounds which must be always ready for use in his mind, and at the same time he will be made familiar with the character of the methodical processes which secure a prompt and sure testimony to the elementary composition of the substances he examines.

4. The subject is treated in two parts or divisions, of which the first contains a systematic course of examination for substances in solution, when once that solution has been made ; and the second treats chiefly of the preliminary examination of solids and the means of bringing them into solution.

PART FIRST.

CHAPTER I.

DIVISION OF THE METALLIC ELEMENTS INTO CLASSES.

5. *Example of the Separation of two Elements.* Put a small crystal of nitrate of silver and a small crystal of sulphate of copper into a test-tube (§ 155), and dissolve them in two teaspoonfuls of water, warming the water at the lamp to facilitate the solution. Add to this solution a few drops of dilute chlorhydric acid (§ 98). Shake the contents of the tube violently, wait until the curdy precipitate, which the acid produces, has separated from the liquid, and then add one more drop of chlorhydric acid. If this drop produces an additional precipitate, repeat the operation until the new drop of acid produces no change in the partially clarified liquid. Then, and not till then, has *all* the silver which the original solution contained been precipitated in the form of chloride of silver, an unemployed balance or *excess* of the *reagent,* chlorhydric acid, remaining in the clear liquid ; this liquid can be readily separated by filtration from the curdy chloride. Shake the contents of the test-tube, and transfer them as completely as possible to a filter (§ 160), supported in a very small glass funnel (§ 159), which has been placed in the mouth of a test-tube. With the wash-bottle (§ 170) rinse into the filter that

portion of the precipitate which has adhered to the sides of the first test-tube. When the filtrate has drained completely from the precipitate, set the test-tube which has received it aside. Wash the precipitate together into the apex of the filter by means of a wash-bottle with a fine outlet; and, in order to wash out the soluble sulphate of copper which adheres to the precipitate, fill the filter full of water two or three times, throwing away this wash-water when it has passed through the filter.

The complete separation of the silver and copper which were mixed in the original solution is already accomplished; the silver is on the filter in the form of chloride; the copper is in the clear, bluish filtrate. This speedy and effectual separation of the two elements is based upon the fact that chloride of silver is insoluble, while chloride of copper is soluble, in water and acid liquids. Such differences of solubility are the chief reliance of the analyst.

6. *Definition of the term "Class." Class I.* In this experiment only two elements have been separated. It might obviously be very difficult, if not impossible, to find a special reagent for every element, which would always precipitate that single element and never any other. Chlorhydric acid, for example, which precipitates silver so admirably from any solution containing that element, is capable of eliminating two other elements under like conditions. The lower chloride of mercury (calomel) is insoluble in water and weak acids. Chloride of lead is sparingly soluble in cold water, and is still less soluble in water acidulated with chlorhydric acid. The chlorides of the other metallic elements are all soluble in water and acids under the conditions of the analytical process.

There are embraced in the scope of this manual twenty-two common elements of the sort ordinarily called *metallic*, most of which form those oxides relatively poor in oxygen which are collectively designated as *bases*. If chlorhydric acid were added in proper quantity to a solution imagined to contain all these elements, three, and only three, of the twenty-two elements would be precipitated as chlorides. After filtration and washing, a mixture of chloride of silver, chloride of lead, and subchloride of mercury, would remain upon the filter, and all the other elements would have passed into the filtrate. Silver, lead, and mercury, the three elements thus separated from the rest by this well-marked reaction with chlorhydric acid, constitute a *class*, the first of several classes into which the metallic elements are divided for the ends of qualitative analysis. Each class is characterized by some clear reaction which suffices, when intelligently applied, to separate the members of any one class from the other classes. The chemical agent, by means of which this distinctive reaction is exhibited, is called the *general reagent* of the class. Thus, chlorhydric acid is the general reagent of the first class.

This division of the elements into classes renders it unnecessary to find means of separating each individual element from *all* the others. In the systematic course of an analysis, the classes are first sought for and separated; afterwards each class is treated by itself for the detection of its individual members. It is an incidental advantage of this division of the elements into classes, that, when the absence of any whole class has been proved by the failure of its peculiar general reagent to produce a precipitate, it is unnecessary to search farther for any member of that class. Much time is thus saved; for it is as easy to prove the absence of a class as of a single element. The full treatment of the first class of elements, comprising, as we have seen, silver, lead, and mercury, is the subject of Chapter II.

7. *Experiment to illustrate the Division of the Metallic Elements into Classes.* We proceed to demonstrate experimentally the chemical facts upon which rests the division of the other metallic elements into convenient classes.

Prepare a complex solution, by mixing together in a small beaker (§ 158) the following solutions, viz. :—a solution of chloride of copper ($CuCl_2$) prepared by dissolving a few grains of oxide of copper in chlorhydric acid ; a solution of arsenious acid in chlorhydric acid ; a solution of ferrous chloride prepared by dissolving a little fine iron wire or filings in chlorhydric acid ; an aqueous solution of chloride of zinc ; an aqueous solution of chloride of calcium ; an aqueous solution of chloride of magnesium ; an aqueous solution of chloride of sodium. If the solutions be all moderately strong, a small teaspoonful of each solution will be enough. Dilute the mixture thus prepared with its own bulk of water. Should any turbidity or precipitate appear, add chlorhydric acid, little by little, until the solution becomes clear. This solution is representative ; it contains at least one member of each of the classes of elements which remain to be defined. It contains no member of the first class, which may be consistently supposed to have been previously precipitated, as in the foregoing experiment (§ 5), an excess of chlorhydric acid remaining in the liquid.

8. *Definition of Classes II and III.* Pass a slow current of sulphydric acid gas (§ 107) from a gas-bottle or generator through the acid liquid in the beaker. This operation must be performed under a hood. A dense, dark-

colored precipitate will immediately appear, and gradually increase in bulk. When the gas has flowed continually for five or ten minutes through the liquid, stop the stream, stir the liquid well, and blow out the sulphydric acid which lies in the beaker. If after the lapse of two or three minutes the liquid smells distinctly of sulphydric acid, it is saturated with the gas, and it is sure that the reagent has done its work. If the liquid does not retain the characteristic odor, the gas must be again passed through it, until saturation is certainly attained.

Pour the contents of the beaker, well stirred up, upon a filter, supported over a test-tube or second beaker. Rinse the first beaker once with a teaspoonful of water, and transfer this rinsing water to the filter, allowing the filtered liquid to mix with the original filtrate. Label* this filtrate "Filtrate from II and III" (classes), and preserve it for later study.

If any considerable quantity of precipitate has adhered to the sides of the original beaker, it may be detached and washed into the filter by means of a sharp jet of water from the wash-bottle. The precipitate, as it lies upon the filter, must then be washed once or twice with water; the wash-water is thrown away. The washed precipitate consists of a mixture of sulphide of copper (CuS) and ter-

---

* The student should at once make it a rule to label every filtrate or precipitate which he has occasion to set aside, even for a few moments. A bit of paper large enough to carry a descriptive symbol or abbreviation should be attached to the vessel which contains the liquid or precipitate. Paper gummed on the back is convenient for this use.

This habit, once acquired, will enable the student to carry on simultaneously, without error or confusion, several operations. He may be throwing down one precipitate, washing another, filtering a third, and dissolving a fourth, at the same time, and the four processes may belong to as many different stages of the analysis. There will be no danger of error, if labels are faithfully used; and a great deal of time will be saved. The unaided memory is incapable of doing such work with that full certainty, admitting of no suspicion or after-qualms of doubt, which is alone satisfying, or indeed, admissible, in scientific research.

sulphide of arsenic (As_2S_3). The fact that these sulphides are precipitated under the conditions of this experiment proves that they are both insoluble in weak acid liquors. They are also both insoluble in water. But an important difference between the two sulphides nevertheless exists, a difference which affords a trustworthy means of separating one from the other.

When the water has drained away from the precipitate, open the filter upon a plate of glass, and gently scrape the precipitate off the paper with a spatula of wood or horn. Place the precipitate in a small porcelain dish (§ 162), pour over it enough sulphydrate of sodium solution (§ 117) to somewhat more than cover it, and heat the mixture cautiously to boiling, stirring it all the time with a glass rod. The quantity of sulphydrate of sodium solution to be employed varies, of course, with the bulk of the precipitate ; but in this case two or three teaspoonfuls will probably suffice. It is very undesirable to use an unnecessarily large quantity of the reagent. A portion of the original precipitate remains undissolved ; but a portion has passed into solution. Filter the hot liquid again. The black residue on the filter is sulphide of copper, which is insoluble, not only in water and weak acids, but also in alkaline liquids. To the filtrate, collected in a test-tube, add gradually chlorhydric acid, until the liquid has an acid reaction on litmus paper (§ 148). A yellow precipitate of sulphide of arsenic will appear as soon as the alkaline solvent which kept it in solution is destroyed. The sulphide of arsenic differs from the sulphide of copper in that it is soluble in alkaline hydrates and sulphides.

But in this series of experiments copper and arsenic stand as representatives of classes. The following common elements have sulphides which are insoluble in water, weak acids, *and* alkaline liquids :—Lead, mercury, bismuth, cadmium and copper. These elements constitute Class II in our system of analysis. The following elements have sulphides which are insoluble in water and weak acids, but soluble in alkaline liquids :—Arsenic, antimony, tin (and the precious metals, gold and platinum). These elements constitute Class III. If all the elements of both groups had been present, the analytical process for separating one class from the other would not have been different from that just executed.

The question may naturally suggest itself, how it happens that lead and mercury are included in Class II when they were both precipitated in Class I. The chloride of lead which is thrown down by chlorhydric acid, is not wholly insoluble in water ; hence it happens that the lead is not completely precipitated in Class I. That portion of the lead which has escaped precipitation as chloride in Class I, will be thrown down as sulphide in Class II, for the sulphide of lead is insoluble in water, weak acids, and alkalies. In regard to mercury, it will be remembered that there are two sorts of mercury salts, mercurous salts and mercuric salts. The mercurous chloride (calomel) is insoluble in water ; but the mercuric chloride (corrosive sublimate) is soluble in water. If, therefore, mercury be present in the form of some mercurous salt, it will be separated as chloride in Class I. If, on the contrary, it be present in the form of some mercuric salt, it will be separated in Class II as mercuric sulphide (HgS), for this sulphide is insoluble in water, weak acids, and alkaline liquids. If a mixture of mercurous and mercuric salts be contained in the original solution, mercury will appear both in Class I and Class II.

The treatment of Class II is fully discussed in Chapter III. The separation of Class III and the means of separating the members of the class, each from the others, form the subject of Chapter IV.

9. *Definition of Class IV.* We now return to the study of the filtrate from Classes II and III. Pour the liquid into a small evaporating dish, and boil it gently for five or six minutes, to expel the sulphuretted hydrogen with which the fluid is still charged. To make sure that all the gas is expelled, hold a bit of white paper moistened with a solution of acetate of lead (§ 138) over the boiling liquid; when the paper remains white, all the sulphuretted hydrogen is gone. Next add to the liquid in the dish ten or twelve drops of strong nitric acid (§ 99), and again gently boil the liquid for three or four minutes, in order that all the iron present may be converted into ferric salts. Then pour the liquid into a test-tube, add to it about one-third its bulk of chloride of ammonium (§ 112), and finally add ammonia-water (§ 109), little by little, until the mixture, after being well shaken, smells decidedly of ammonia. A brownish red precipitate of hydrated sesquioxide of iron will separate from the liquid. Ammonia-water precipitates this familiar iron compound, *even in the presence of salts of ammonium,* such as the chloride of ammonium which has been expressly added, and the nitrate of ammonium which has been formed during the neutralization of the acid liquid. Pour the contents of the test-tube upon a filter, rinse the tube and the precipitate once with a little water, and preserve the whole filtrate for subsequent operations.

Aluminum and chromium are precipitated, as iron has here been, by ammonia-water under the same conditions and in the same form, namely, as hydrates. These three elements, therefore, constitute the fourth class, whose treatment forms the subject of Chapter V. The student may be curious to know why the presence of ammonium-salts is insisted upon before the elements of this class are thrown down by ammonia-water. The ammonium-salts have nothing to do with the precipitation of iron, aluminum and chromium; but by their faculty of forming soluble double salts, they prevent the partial precipitation of certain other elements which are more conveniently dealt with in classes which are to follow. The ammonium-salts keep in solution certain other elements which otherwise would encumber Class IV.

10. *Definition of Class V.* We now proceed to the examination of the filtrate from the precipitate of Class IV. Bring this liquid to boiling in a test-tube (or in a small flask, if there be much liquid), and add sulphydrate of ammonium (§ 110), little by little, to the boiling liquid, as long as a precipitate continues to be formed. To make sure that the precipitation is complete, shake the hot contents of the test-tube strongly, and then allow the mixture to settle until the upper portion of the liquid becomes clear. Into this clear portion let fall a drop of sulphydrate of ammonium ; when this drop produces no additional precipitate, the precipitation is complete. Filter off the whitish precipitate of sulphide of zinc, and preserve the filtrate for further treatment.

The element zinc, representing a new class of elements, is precipitated under the conditions of the above experiment, because its sulphide, though soluble in dilute acids, is insoluble in alkaline liquids. The metals manganese, nickel, and cobalt resemble zinc in this respect, and these four elements, therefore, form a new class, Class V, in this analytical method. The representative sulphide of this class was not precipitated by the sulphydric acid when that reagent was employed to throw down the members of Classes II and III, because the solution was at that time acid. Again, it was not precipitated with Class IV by the ammonia-water, because the sulphydric acid gas, with which the solution had previously been charged, was expelled by boiling before the ammonia-water was added. The complete treatment of Class V forms the subject of Chapter VI.

11. *Definition of Class VI.* Add to the filtrate, from Class V, two or three teaspoonfuls of carbonate of ammonium (§ 111), and boil the solution. A white precipitate of carbonate of calcium will be produced. After boiling, allow the precipitate to settle until the upper portion of the liquid is comparatively clear. To this clarified portion add a fresh drop of carbonate of ammonium. If this drop produce an additional precipitate, more carbonate of ammonium must be added, and the boiling repeated. To the partially clarified liquid add again a drop of carbonate of ammonium. This process of making sure of the complete precipitation of the calcium is essentially the same as that prescribed in precipitating the last class, and is, indeed, of general application. When the precipitation of the calcium has been proved to be complete, filter the whole liquid, and receive the filtrate in a small evaporating dish. Calcium is separated in the form of carbonate under these circumstances, because this carbonate is almost insoluble in weak alkaline liquids, when an excess of carbonate of ammonium is present. The allied elements barium and strontium behave in the same way, so that these three elements, viz., barium, strontium, and calcium, compose a new class—Class VI, whose complete treatment is set forth in Chapter VII.

12. *Definition of Class VII.* Of the twenty-two metallic elements, which were to be classified (§ 6), only three remain, viz., magnesium, sodium, and potassium. It is obvious that these three elements could not have remained in solution through all the operations to which the original liquid has been submitted, unless their chlorides and sulphides had been soluble in weak acids, and their oxides, sulphides, and carbonates, soluble in dilute ammonia-water, at least in presence of dilute solutions of ammonium-salts. It is a fact that all these compounds of sodium and potassium are soluble in water, and in weak acid, alkaline, and saline solutions; the magnesium would have been partially precipitated in Classes IV, V, and VI, but for the presence of ammonium-salts in the solution. These three elements constitute Class VII.

Evaporate the filtrate from Class VI until it is reduced to one-half or one-third of its original bulk. Pour a small part of the evaporated filtrate into a test-tube; add a little ammonia-water and a teaspoonful of phosphate of sodium (§ 121), and shake the contents of the tube violently. Sooner or later a crystalline precipitate will appear. This peculiar white precipitate of phosphate of magnesium and ammonium identifies magnesium; but as we have added a reagent containing sodium, the filtrate is useless for further examination. The liquid remaining in the evaporating-dish is then evaporated to dryness, and moderately ignited until fuming ceases. All the ammoniacal salts which the solution contained will be driven off by this means, and there will remain a fixed residue, in which are concentrated all the salts of magnesium and sodium which the original solution contained. In this case we have proved the presence of magnesium; it remains to indicate briefly the nature of the means used to detect sodium.

Dissolve the residue in the dish, or a portion of it, in three or four drops of water. Dip a clean platinum wire

(§ 168) into this solution, and introduce this wire into the colorless flame of a gas or spirit-lamp (§ 163). An intense yellow coloration of the flame demonstrates the presence of sodium. A violent coloration would have proved the presence of potassium. Magnesium compounds, when present, have no prejudicial effect on these characteristic colorations. The means of detecting each member of this last class in presence of the others will be found described in Chapter VIII.

13. A condensed statement of the classification illustrated by the foregoing experiments is contained in the table on the next page. All the common metallic elements are embraced in it. The place of the precious metals gold and platinum is also indicated. The classification itself would not be essentially different, if all the rare elements were comprehended in it. The general subdivisions would be the same, although some of them would embrace many more particulars.

14. It is essential to success to follow precisely the prescribed *order* in applying the various general reagents. Class I would go down with Class II, were chlorhydric acid forgotten as the first general reagent. Class II would be precipitated in part with Class IV and in part with Class V, if sulphuretted hydrogen should not be used in its proper place. A large number of the members of the first five classes would be precipitated as carbonates with Class VI, were they not previously eliminated by the systematic application of chlorhydric acid, sulphuretted hydrogen, ammonia-water, and sulphide of ammonium, in the precise order and under the exact conditions above described. It should be noticed that all the general reagents are volatile substances, which can be completely

THE SEVEN CLASSES OF THE METALLIC ELEMENTS.

CLASS I.	CLASS II.	CLASS III.	CLASS IV.	CLASS V.	CLASS VI.	CLASS VII.
Precipitated as chlorides,	Precipitated as sulphides insoluble in dilute acids, and not redissolvable by alkaline liquids,	Precipitated as sulphides insoluble in dilute acids, but redissolved by alkaline liquids,	Precipitated by ammonia water, usually as hydrates,	Precipitated as sulphides insoluble in alkaline fluids,	Precipitated as carbonates,	Remaining elements. Distinguished by special tests,
Ag Pb Hg	Pb Hg Bi Cd Cu	As Sb Sn [Au] [Pt]	Fe Al Cr [together with certain salts which require an acid solvent.]	Zn Mn Ni Co	Ca Ba Sr	Mg Na K

removed by an evaporation to dryness followed by a very moderate ignition.

15. The series of experiments just completed is merely intended to demonstrate the principles in accordance with which these twenty-two common elements are classified for the purposes of qualitative analysis. The general plan is here sketched. The practical details, essential to success in the conduct of an actual analysis, will be given hereafter.

CHAPTER II.

CLASS I. CHLORIDES INSOLUBLE IN WATER AND ACIDS.

16. *Example of the Precipitation of the Members of Class I.* Place in a test-tube five or six drops of a tolerably concentrated aqueous solution of nitrate of silver, an equal quantity of a solution of mercurous nitrate, and two teaspoonfuls of a solution of nitrate of lead. In case the solution becomes turbid through the action of carbonic acid dissolved in the water, pour in one or two drops of nitric acid to destroy the cloudiness.

Add dilute chlorhydric acid to the solution, drop by drop, and shake the mixture thoroughly after each addition of the acid, until the fresh portions of the latter cease to form any precipitate on coming in contact with the comparatively clear liquor which floats above the insoluble chlorides. Finally add three or four more drops of the acid to insure the presence of an excess of it in the solution.

17. *Analysis of the Mixed Chlorides.* The following method of separating the chlorides of lead, silver, and mercury, one from another, depends upon the facts :—1st. That chloride of lead, though but little soluble in cold water, dissolves readily in boiling water, while chloride of silver and subchloride of mercury are as good as insoluble in that liquid. 2d. That chloride of silver is soluble in ammonia-water; and 3d. That subchloride of mercury is discolored by ammonia-water without dissolving.

To effect the separation :—Collect the precipitate, produced by chlorhydric acid, upon a filter, allow it to drain, and rinse it with a few drops of cold water. Place a clean test-tube beneath the funnel which contains the filter and precipitate, thrust a glass rod through the apex of the filter, and wash the precipitate off the filter into the test-tube by means of a wash-bottle which throws a fine stream.

Heat the mixture of water and precipitate to boiling, pour the hot liquor, but not the precipitate, upon a new filter, and add to the filtrate two or three teaspoonfuls of dilute sulphuric acid (§ 103). A white cloud of sulphate of lead will be formed in the midst of the liquid. In case the precipitate contains a large proportion of chloride of lead, it may happen that the hot water will take up so much of it that crystals of the chloride will separate from the clear aqueous solution as it becomes cold, or that the liquor will be rendered cloudy by the deposition of numerous small particles of the chloride.

Test for **Pb.**

After the mixed precipitate of chloride of silver and subchloride of mercury which was left in the test-tube has been boiled several times with fresh water, in order to insure the removal of all the chloride of lead, transfer the whole precipitate to the filter of the last paragraph, and pour some ammonia-water which has been heated in another test-tube, directly upon the white precipitate as it lies on the filter, placing a clean test-tube beneath the funnel. The chloride of silver, dissolved by the ammonia-water, will pass into the filtrate, while the subchloride of mercury suffers decomposition, and is converted into an obscure compound of mercury, chlorine, nitrogen, and hydrogen, which remains upon the filter in the form of an insoluble black or gray powder.

To confirm the presence of silver, neutralize the ammoniacal solution with dilute nitric acid (§ 100), and observe that the chloride of silver is reprecipitated.

To confirm the presence of mercury, the metal itself may be set free by heating the dry residue with carbonate of sodium (§ 118) in a glass tube. To insure the success of this experiment, wash into the lowest point of the filter the whole of the black residue, including those portions which have remained adhering to the sides of the test-tube. As soon as the last drops of liquid have drained from the filter, dry the latter, either in a dish upon a water-bath, or by spreading it open upon a ring of the iron stand (§ 165) placed high above a small flame of the gas-lamp. When the precipitate is completely dry, Test for Hg. scrape it from the paper, mix it with an equal bulk of carbonate of sodium, previously dried over the gas-lamp on platinum foil (§ 168), and transfer the mixture to the bottom of a glass tube, No. 6 (§ 171), closed at one end. Then wipe out the inside of the tube with a tuft of cotton fixed to a wire, or with a twisted strip of paper, and heat the closed end of the tube for two or three minutes in the flame of a gas-lamp. A sublimate of finely divided metallic mercury will form upon the walls of the tube ; the metal will cohere to visible globules when scratched with a piece of iron wire.

18. An outline of the operations described in the foregoing paragraphs may be presented in tabular form, as follows :—

<table>
<tr><td colspan="3">The General Reagent (HCl) of Class I precipitates $PbCl_2$, AgCl and HgCl. The precipitate is boiled with water :—</td></tr>
<tr><td rowspan="2">$PbCl_2$ goes into solution. Confirm presence of lead by precipitation of sulphate of lead.</td><td colspan="2">AgCl and HgCl remain undissolved. On treating the mixture with ammonia-water,</td></tr>
<tr><td>AgCl dissolves. Confirm presence of silver with nitric acid.</td><td>A black compound of mercury remains undissolved. Confirm presence of mercury by isolating the metal.</td></tr>
</table>

CHAPTER III.

CLASS II. SULPHIDES INSOLUBLE IN WATER, DILUTE ACIDS, AND ALKALIES.

19. *Example of the Precipitation of the Members of Class II.* Place in a small beaker six or eight drops of a strong solution of each of the following substances :—Chloride of mercury (corrosive sublimate), chloride or nitrate of bismuth, of cadmium, and of copper, together with two or three teaspoonfuls of a cold aqueous solution of chloride of lead. Fill the beaker half full of water, and add, drop by drop, enough strong chlorhydric acid to redissolve the basic chloride of bismuth which the water precipitates.

Place the beaker beneath a chimney or in a strong draught of air, and saturate the solution with sulphuretted hydrogen gas. To determine when enough sulphuretted hydrogen has been passed through the liquid, remove the beaker every four or five minutes from the source of the gas, blow away the gas which lies in the beaker above the liquid, and stir the latter thoroughly with a glass rod. If after the lapse of two or three minutes, the liquid still smells strongly of sulphuretted hydrogen, it is saturated with the gas and ready to be filtered. But in case no persistent odor of sulphuretted hydrogen is observed, the gas must be passed anew through the liquor until it has become fully saturated. Since some of the substances above enumerated are thrown down more quickly by sulphuretted hydrogen than the others, it is absolutely necessary to employ the reagent *in excess* in order that those members of the class which are least easily precipitated may not escape detection.

20. *Analysis of the Mixed Sulphides.* The following method of separating the members of Class II depends upon the facts :—1st. That mercuric sulphide is insoluble in dilute nitric acid, while the other sulphides are soluble therein. 2d. That sulphate of lead is insoluble in acidulated water, while the sulphates of the other members of the class are soluble. 3d. That hydrate of bismuth is insoluble in ammonia-water, while the hydrates of cadmium and copper are soluble in that liquid. 4th. That sulphide of cadmium is soluble, and sulphide of copper insoluble, in hot dilute sulphuric acid.

To effect the separation :—Collect the precipitated sulphides upon a filter ; wash the precipitate *thoroughly* with water, in order to completely remove the chlorhydric acid which adheres to it ; make a hole in the filter, and rinse the precipitate into a small porcelain dish. Carefully decant the water from the precipitate, pour upon the latter four or five times as much dilute nitric acid as would be sufficient to cover it, and boil the mixture during five minutes, stirring it constantly with a glass rod, and adding dilute nitric acid at intervals to replace the liquid which evaporates. A heavy dark mass of sulphide of mercury, mixed with some free sulphur resulting from the decomposition of the other sulphides, will remain undissolved. Decant the nitric acid solution into a filter, collect the filtrate in a second porcelain dish, and evaporate it nearly to dryness.

Pour water upon the residue insoluble in nitric acid which was left in the first dish, to remove the adhering liquid, and after this wash-water has been decanted, boil the residue with as much aqua regia (§ 101) as will barely cover it. Dilute the acid solution with an equal volume of water, remove from it, by filtration or otherwise, any particles of free sulphur which may remain undissolved, and add to it almost, but not quite, enough ammonia-water to neutralize its acidity. Place in the still slightly acid solution a small bit of bright copper wire, warm the liquor, and observe that metallic mercury is deposited upon the copper as a white silvery coating. After the lapse of ten or fifteen minutes, dry the wire upon blotting paper, drop it into a narrow glass tube which has been sealed at one end, and heat it at the lamp. Metallic mercury will sublime, and be deposited as a dull mirror upon the cold portions of the glass. By scratching the sublimate with the point of a bit of iron wire, the metal may be made to collect into visible globules.

Test for Hg.

When the greater part of the free nitric acid has been driven off from the filtrate which contains the mixed nitrates of lead, bismuth, cadmium and copper, transfer the residual liquor to a test-tube, mix it with two or three times its volume of dilute sulphuric acid, and leave the mixture at rest during fifteen or twenty minutes. Sulphate of lead will be thrown down as a heavy white powder, plainly to be seen in the test-tube, though it would have been scarcely visible in the white dish.

To confirm the presence of lead, collect the precipitate upon a filter, wash it with water, transfer it to a test-tube, pour upon it two or three times its volume of a solution of normal chromate of potassium (§ 124), and heat the mixture to boiling. The white sulphate of lead will be converted into yellow chromate of lead, without dissolving, and the characteristic color of the latter may be made manifest by collecting it upon a filter, and washing it with water until the excess of chromate of potassium has been completely removed.

Collect the filtrate from the sulphate of lead in a small beaker, and add to it ammonia-water by repeated small portions, taking care to stir the liquid thoroughly after each addition of the ammonia, until a strong persistent odor of the latter is perceptible. The hydrates of copper, cadmium, and bismuth, will all be thrown down at first, but the hydrates of copper and cadmium will redissolve in the excess of ammonia-water, and hydrate of bismuth will alone be left as an insoluble precipitate.

Test for **Bi.**

To prove that this precipitate contains bismuth:—Collect it upon a filter, allow it to drain, and dissolve it in the smallest possible quantity of strong chlorhydric acid, poured drop by drop upon the sides of the filter; carefully evaporate the acid solution to the bulk of two or three drops and pour it into a large test-tube nearly full of water. A dense milky cloud of insoluble basic chloride of bismuth will appear in the water. Since sulphate of lead is not absolutely insoluble in water which contains nitric acid, a slight precipitate of hydrate of lead might be produced on the addition of the ammonia-water even when no bismuth was present in the solution. To prove the presence of bismuth, the oxychloride must always be carefully tested for.

The blue color of the ammoniacal filtrate from the hydrate of bismuth indicates the presence of copper, and when well defined is of itself a sufficient proof of the presence of this element. But in the absence of a marked blue coloration at this stage, copper should be specially tested for in the manner described below.

To separate the cadmium from the copper, proceed as follows :—Transfer the ammoniacal filtrate to a glass flask, heat it to boiling, and drop into the boiling liquid sulphydrate of ammonium as long as a precipitate continues to be formed. In order to be sure that the precipitation is complete, remove the flask from the lamp at intervals, shake it strongly, and allow its contents to settle, so that a comparatively clear liquid may appear at the top, and into this clear liquid pour a drop of the sulphydrate.

As a general rule, the operations of boiling and agitating tend to increase the coherency of precipitates, and to render them in some sense granular, so that they separate completely from the liquid in which they form, leaving it clear and susceptible of rapid filtration.

Collect the precipitate upon a filter, rinse it once or twice with water, and allow it to drain ; then open the filter on a plate of glass, scrape the precipitate off the paper and put it in a porcelain dish. Prepare a dilute sulphuric acid by mixing 1 part by measure of the strong acid (§ 103), with 5 parts of water. Cover the precipitate in the dish liberally with this dilute acid, and heat the mixture until it actually boils ; then pour the boiling liquor upon a filter, and collect the clear filtrate in a beaker. Sulphide of cadmium alone will dissolve in hot dilute acid of the prescribed strength, the black sulphide of copper remaining intact.

Test for Cd.

To prove the presence of cadmium, pass sulphuretted hydrogen gas into the acid filtrate, and observe that the liquor immediately becomes cloudy from the presence of minute particles of sulphide of cadmium of characteristic yellow color. After some time this precipitate will collect at the bottom of the liquid.

Test for Cu.

To prove the presence of copper, in case no blue coloration was visible in the filtrate from the hydrate of bismuth, transfer the black precipitate, insoluble in dilute sulphuric acid, to an evaporating dish, dissolve it in a few drops of boiling, concentrated nitric acid, remove and wash the the spongy mass of sulphur which is set free, neutralize the nitric acid with ammonia-water, acidify the solution with acetic acid (§ 105), transfer it to a test-tube, and add one or two drops of a solution of ferrocyanide of potassium (§ 125). A peculiar reddish-brown precipitate of ferrocyanide of copper will fall in case much copper be present, and even when the proportion of copper in the solution is extremely small, a light brownish-red cloudiness will be produced.

In the absence of copper, yellow sulphide of cadmium would at once be thrown down by the sulphydrate of ammonium, when this reagent is added to the filtrate from the oxide of bismuth; and no further evidence of the presence of cadmium would be required.

21. The operations above described may be presented in tabular form, as follows :—

<table>
<tr><td colspan="5">The General Reagent (H_2S) of Class II precipitates HgS, PbS, Bi_2S_3, CdS, and CuS (as well as members of Class III, which are subsequently separated by solution in sulphydrate of sodium). The precipitate is boiled with nitric acid:—</td></tr>
<tr><td rowspan="4">A residue of HgS, plus S, remains. Confirm presence of mercury with copper wire.</td><td colspan="4">The nitrates of Pb, Bi, Cd, and Cu go into solution. On adding dilute sulphuric acid to the concentrated solution:—</td></tr>
<tr><td rowspan="3">$PbSO_4$, is thrown down. Confirm the presence of Pb by converting $PbSO_4$ into $PbCrO_4$.</td><td colspan="3">The sulphates of Bi, Cd, and Cu remain in solution. On adding an excess of ammonia-water:—</td></tr>
<tr><td rowspan="2">Hydrate of Bismuth is thrown down. Confirm Bi by precipitating the oxychloride.</td><td colspan="2">Compounds of Cd and of Cu remain in solution. Throw down CdS and CuS with $(NH_4)HS$, and boil with dilute sulphuric acid:—</td></tr>
<tr><td>$CdSO_4$ goes into solution. Confirm presence of Cd by precipitation of CdS.</td><td>CuS remains undissolved. Confirm presence of Cu by testing with ferrocyanide of potassium.</td></tr>
</table>

22. *The method of Separating Class I from Class II* has already been particularly described in § 6. It should be observed, however, that even if no member of Class I were present in the mixture to be analyzed, it would still be necessary to acidulate the liquid with chlorhydric acid, before passing the sulphuretted hydrogen, in order to prevent the precipitation of members of Classes IV and V, and to secure the complete precipitation of the members of Class III.

The liquid should be watched attentively when the stream of sulphuretted hydrogen first begins to flow through it, since useful inferences may often be drawn from the various phenomena which present themselves.

a. Thus, the formation of a white precipitate which

afterward changes to yellow, orange, brownish-red, and finally to black, as the liquid gradually becomes saturated with the gas, indicates the presence of mercuric chloride. The white precipitate at first formed is a compound of chloride and sulphide of mercury ($HgCl_2$; $2HgS$), but by the action of successive portions of sulphuretted hydrogen, the composition and appearance of the precipitate is changed, until it has been completely converted into black sulphide of mercury.

b. If the precipitate is of a dull red color at first, afterward changing to black, the probable presence of lead is indicated ; for sulphuretted hydrogen throws down from solutions which contain much free chlorhydric acid a red compound of chloride of lead and sulphide of lead, which is afterward decomposed, with formation of the black sulphide, when the solution becomes saturated with the gas.

c. A decided bright yellow precipitate would indicate the presence of cadmium, arsenic, or tin ; of these three, cadmium is distinguished by the fact that its sulphide remains undissolved when the precipitate is treated with sulphydrate of sodium to separate the members of Class III.

But even from solutions which contain no members of Classes II or III, yellowish-white or milky-white precipitates of free sulphur are often thrown down ; for sulphuretted hydrogen is easily decomposed, with deposition of sulphur, by a variety of oxidizing agents, such as nitric, chromic, and chloric acids, and solutions of ferric salts and of free chlorine. If the solution under examination contained much nitric acid, sulphuretted hydrogen would have to be passed through it for a long time to destroy the acid, before the liquid could be saturated with the gas. In this case the sulphur separates as a tenacious mass of dirty yellow color ; but in most instances, notably when the solution contains a ferric salt, the sulphur is precipitated in the form of exceedingly minute particles, which impart to the solution a peculiar milkiness or opalescence. These particles are so fine that they pass through the pores of filter paper ; they cannot be completely removed by filtration.

If the original solution contains a chromate, its yellow or reddish-yellow color will be changed to green by the action of sulphuretted hydrogen ; for the chromate is reduced to the condition of sesquichloride of chromium :—

$$2MCrO_4+10HCl+3H_2S=Cr_2Cl_6+3S+2MCl_2+8H_2O.$$

The sulphur is set free in the form of the minute white particles above described, and remains suspended in the green liquid, looking not unlike a green precipitate.

A white precipitate of sulphur would be thrown down even before the passage of sulphuretted hydrogen, in case the original solution contained a hyposulphite ; for this class of salts is decomposed, with evolution of sulphurous acid and deposition of sulphur, on the addition of the general reagent (HCl) of Class I. Some sulphides also are decomposed by chlorhydric acid, with deposition of sulphur. A gelatinous white precipitate of hydrated silicic acid might also be formed at this stage in certain circumstances, as will be explained hereafter (§§ 67, 80, *a*).

d. The immediate formation of a black precipitate indicates the presence of copper or bismuth, and it is to be observed that either of these black precipitates would obscure the colors of the other sulphides of the class, and conceal them if present.

e. If no precipitate appears even when the liquid has become saturated with the gas, the absence of every member of Classes II and III is, of course, to be inferred.

CHAPTER IV.

CLASS III. SULFHIDES INSOLUBLE IN WATER OR DILUTE ACIDS, BUT SOLUBLE IN ALKALINE SOLUTIONS.

23. *Example of the Precipitation of the Members of Class III.* Place in a small beaker six or eight drops of strong solutions of the chlorides of arsenic, antimony, and tin. Pour in enough dilute chlorhydric acid to half fill the beaker, and, if need be, a sufficient number of drops of strong chlorhydric acid to dissolve any cloud of basic chloride of antimony which may appear in the liquor.

Pass sulphuretted hydrogen gas through the solution, in the manner described in § 19, until the odor of the gas persists. Then collect the precipitated sulphides upon a filter, and rinse the precipitate once or twice with water.

In the analysis of any complex solution of unknown composition which might contain one or all of the members of Class III, the sulphides of this class would, of course, all be thrown down at the same time as those of Class II. (Compare §§ 8, 21.) It will be well, therefore, for the sake of illustration, for the student to dissolve the present precipitate in sulphydrate of sodium, in order that he may begin the treatment of Class III at the precise point at which this class would be encountered in an actual analysis—namely, with the sulphides of the class in alkaline solution. To this end, allow the washed precipitate to drain, spread out the filter upon a plate of glass, scrape the precipitate from the paper with a small spatula of platinum, horn, or wood, and transfer it to a porcelain dish. Pour upon the precipitate two or three times as much of a solution of sulphydrate of sodium as would be sufficient to cover it, and boil the mixture very cautiously, so as to avoid spattering. The precipitate will soon dissolve, and no solid matter will be left suspended in the solution, except a few fibres of the filter paper. It is such a solution as this which in an actual analysis is examined for members of Class III.

Pour the alkaline solution into a beaker, and stir into it, little by little, dilute chlorhydric acid until the liquid exhibits an acid reaction. The sulphides of arsenic, antimony and tin, are re-precipitated, as such, together with a quantity of sulphur derived from the decomposition of the sulphydrate of sodium.

24. *Analysis of the Mixed Sulphides.* The method here given of separating arsenic, antimony and tin, depends :— 1st. Upon the solubility of sulphide of arsenic in a dilute aqueous solution of carbonate of ammonium, and the insolubility, or very slight solubility of the sulphides of antimony and tin in that liquid ; 2d. Upon the oxidation of the sulphides of antimony and tin by fusion with nitrate of sodium ; 3d. Upon the reduction of the compounds thus formed to the metallic state by zinc ; 4th. Upon the solubility of tin in chlorhydric acid.

To effect the separation :—Collect upon a filter the precipitate produced by acidifying the sulphydrate of sodium solution, wash it with water to remove the acid and the chloride of sodium which adhere to it, and allow it to drain. Spread out the filter in a small porcelain dish, and cover its contents with a dilute aqueous solution of carbonate of ammonium—obtained by dissolving 1 part of the solid carbonate in 12 parts of water, or by mixing one volume of the strong solution (§ 111) employed as the general reagent of Class VI, with two volumes of water—and stir the mixture. After the lapse of four or five minutes, pour the carbonate of ammonium solution upon a new filter, collect the filtrate in a beaker, and stir into it successive drops of dilute chlorhydric acid—taking care to add no more than a single drop of the acid at any one time—until carbonic acid ceases to escape, and the liquid exhibits a strong acid reaction when tested with litmus paper. A bright-yellow precipitate of sulphide of arsenic will separate immediately from the acid liquor, in case much arsenic is present ; or a yellow cloudiness will appear at first, if the quantity of arsenic in the solution be minute. In the latter case the mixture must be left at rest for some hours, in order that distinct yellow flocks of the sulphide may collect at the bottom of the vessel.

To confirm the presence of arsenic:—Collect the precipitated sulphide upon a filter, dry it *thoroughly* at a gentle heat, scrape it from the paper, place it in a narrow glass tube which has been blown to a bulb at one end (see § 175), and cover it with five or six times its bulk of a *perfectly dry* mixture of equal parts of carbonate of sodium and cyanide of potassium (§ 127). The bulb of the tube must be large enough to hold twice as much of the mixture as is really to be placed in it, in order that there may be room for the mass to swell when it is heated and fused. After the bulb has been charged, wipe out the inside of the tube with a tuft of cotton fixed to a wire, or with a twisted strip of paper, and heat the contents of the bulb during two or three minutes in the flame of a gas-lamp. A dark, lustrous ring of metallic arsenic will be deposited upon the cold walls of the tube.

Test for As.

The mixed precipitate of sulphide of antimony and sulphide of tin, insoluble in carbonate of ammonium, is treated as follows :—Scrape and wash all the precipitate off the paper into the dish, and add a teaspoonful of strong nitric acid. Evaporate the mixture to the bulk of a teaspoonful without heeding the mass of sulphur which separates; then add to it two or three teaspoonfuls of a strong solution of nitrate of sodium (§ 120), and evaporate the mixture to dryness with constant stirring. Heat the dry mass until it fuses. When the dish has become cold, fill it half full of cold water; when the water has softened the cake of fused salts, stir the contents of the dish, break up the lumps, and finally filter the insoluble tin and antimony compounds from the soluble nitrates which the water has taken up. This insoluble residue must next be placed in a porcelain dish in contact with a slip of clean, smooth, platinum foil, and heated with a little chlorhydric acid. Water is then added, and a small fragment of zinc is put into the liquid. Tin and antimony will be reduced to the metallic state by the zinc. It is found that a part of the reduced antimony attaches itself firmly to the platinum foil, producing a very characteristic black stain upon the bright surface of the foil. Tin produces no such effect. As soon as the hydrogen has ceased, or nearly ceased, to escape from the liquid, the solution, which consists of chloride of zinc, is carefully poured off. The residue, together with the remnant of zinc if any tin adhere to it, is warmed with strong chlorhydric acid, in which the tin dissolves. Pour off this solution of protochloride of tin, and add to it two or three drops of a solution of mercuric chloride (corrosive sublimate) (§ 145). A white or gray precipitate of mercurous chloride (calomel), often mixed with gray metallic mercury will be thrown down, for:—

Test for **Sb.**

Test for **Sn.**

$$2HgCl_2+SnCl_2=2HgCl+SnCl_4; \text{ and}$$
$$2HgCl+SnCl_2=2Hg+SnCl_4.$$

To prove that the precipitate really contains calomel, decant the supernatant liquid, cover the precipitate with ammonia-water, and heat the mixture to boiling (compare p. 28).

The black stain upon platinum is a trustworthy indication of the presence of antimony ; but a verification of its presence may be readily obtained by the following process :—Wash the residue from the tin solution thoroughly by decantation, and warm it for a quarter of an hour with a solution of tartaric acid (§ 106), adding a few drops of nitric acid towards the close of the digestion. Care must be taken not to heat any part of the dish so hot as to burn or blacken the tartaric acid.

Antimony dissolves in this tartaric acid solution ; its presence may be proved by passing sulphuretted hydrogen through the decanted solution. An orange-red-colored precipitate of sulphide of antimony will be sooner or later thrown down.

25. An outline of the foregoing operations may be represented in tabular form, as follows :—

The General Reagent (H_2S) of Class III precipitates As_2S_3, Sb_2S_3, and SnS or SnS_2, [Au_2S_3 and PtS_2]. (As well as the members of Class II, from which Class III is separated by solution in sulphydrate of sodium and reprecipitation with an acid.) Digest the precipitate, thrown down by acid from the sulphydrate of sodium solution, in a weak solution of carbonate of ammonium :	
As_2S_3 dissolves and may be reprecipitated by neutralizing the solution with HCl. Confirm presence of As by reducing As_2S_3 with $Na_2CO_3 + KCN$.	Sb_2S_3, SnS and SnS_2 [Au_2S_3 and PtS_2], remain undissolved. Treat with HNO_3 ; fuse with $NaNO_3$; treat fused salts with H_2O. Reduce the insoluble residue with zinc and dilute acid in presence of platinum foil. Sb is deposited on the foil. Treat metallic residue with HCl. Sn dissolves ; its presence is proved by $HgCl_2$. Sb is then dissolved in tartaric acid, and reprecipitated as orange Sb_2S_3.

Under some circumstances, acids fail to reprecipitate completely the sulphides which are dissolved in sulphydrate of sodium. This difficulty is particularly likely to occur when tin is present. It may be avoided by modifying the above process, as follows: add strong nitric acid to the sulphydrate of sodium solution to acid reaction, and without regard to the precipitate which separates, evaporate the acid mixture to the bulk of one or two teaspoonfuls. Then add three or four teaspoonfuls of a strong solution of nitrate of sodium, evaporate to dryness with constant stirring, and heat the mass until it fuses. Let the dish cool, and then soak the fused salts in three or four teaspoonfuls of cold water, until they soften and in part dissolve. Break up the lumps, stir repeatedly, and finally filter. Arseniate of sodium, together with some nitrate and sulphate of sodium, passes into solution, while the antimony and tin are to be found in the insoluble residue. This residue is examined for these two metals precisely as above described (§ 24, p. 41).

The aqueous solution is tested for arsenic as follows: Acidulate the solution faintly with nitric acid, and warm it; then add a few drops of a prepared solution of sulphate of magnesium and chloride of ammonium (§ 139), and set the mixture aside for twelve hours. The formation of a white, crystalline precipitate of arseniate of ammonium and magnesium is evidence of the presence of arsenic. Any precipitate supposed to contain arsenic or antimony may be further examined by converting the arsenic or antimony into a hydrogen compound, by the method known as Marsh's test (see the authors' Manual of Inorganic Chemistry, pp. 259, 270). This method is applicable to the detection of very minute quantities of arsenic or antimony.

Test for As.

In the case of mixtures containing gold and platinum (see § 13), as well as arsenic, antimony, and tin, the gold and platinum would remain with the tin, without interfering in any way with the separation or detection of either member of the class. Since the sulphides of gold and platinum are both black, while those of arsenic and tin are yellow or brown, and that of antimony is orange, the presence of any considerable quantity of either of the precious metals would be indicated by the black color of the class-precipitate. There are excellent special tests both for gold and platinum, by which these elements may be detected even in the presence of all the other metals. Hence it is most convenient to make special search for them in the original substance, by methods to be described hereafter (§ 92 *b*), whenever the preliminary examination has given reason to suspect the presence of either of them.

26. *The Method of Separating Class II from Class III* has been sufficiently described in §§ 8, 23. Attention should be paid to the color of the precipitate produced when the sulphydrate of sodium solution is acidified, although the color of this precipitate is obscured by the whitish mass of sulphur derived from the decomposition of the sulphydrate of sodium itself.

An orange-colored precipitate indicates the presence of antimony ;

A bright yellow precipitate, the presence of arsenic ;

A dull yellow precipitate, white at first, the presence of bisulphide of tin ; and

A dark brown precipitate, the presence of protosulphide of tin.

A black precipitate might indicate the presence of gold or platinum, or might be due to the presence of the sulphide of mercury or of copper ; for both these sulphides are somewhat soluble in sulphydrate of sodium in presence of the sulphides of Class III. (Compare § 43.)

Gold and platinum must be specially sought for according to the methods set forth in § 92 *b*.

These colorations are valuable only in so far as they afford evidence of the presence of some one member of the class ; they by no means prove the absence of the other members.

CHAPTER V.

CLASS IV. HYDRATES INSOLUBLE IN WATER, AMMONIA-WATER, AND SOLUTIONS OF AMMONIUM-SALTS.

27. The leading fact upon which the separation of this class is based is the insolubility of the hydrates of iron, aluminum and chromium in ammonia-water, even in presence of solutions of ammonium-salts. But these three hydrates are not the only substances which are liable to be precipitated in an actual analysis when ammonia-water is added in excess to a solution previously acid. There are a number of salts, soluble in acids, but not in water or in weak alkaline liquids, which are thrown down as salts, without change, when their acid solvent is destroyed.

It is clear that it is needless to provide in this place against the presence of such salts of elements belonging to Classes I, II, and III. Those elements are already eliminated when the fourth class is taken in hand. But if there are any salts of elements belonging to the fourth and higher classes which can only be kept in solution by a free acid, they will be precipitated without change in consequence of the neutralization of their solvent by the ammonia-water added to precipitate the three hydrates above mentioned. Such salts are the phosphates of several members of Classes IV, VI, and VII, besides a number of oxalates, borates, silicates, and fluorides, which occur so seldom that they need not be particularly considered in an elementary treatise. Beside the phosphates, several chromites and aluminates of members of Classes VI and VII are insoluble in ammonia-water, and are often thrown down wholly or in part along with the legitimate members of Class IV. Manganese also (a member of Class V) is frequently precipitated in combination with members of Class IV in the form of chromite, ferrite or aluminate of manganese. The general scheme for the examination of Class IV necessarily provides for the detection of all the members of the class in the possible presence of these extraneous substances.

28. *Example of the Precipitation with Ammonia-water.* Pour into a beaker six or eight drops of aqueous solutions of sulphate of manganese, common alum, and chrome-alum, and two or three drops of a solution of sulphate of iron (copperas). Dissolve in a few drops of strong boiling chlorhydric acid as much bone-ash as could be held on half a pea, and add the solution to those already placed in the beaker.

Fill the beaker about one-third full of water, heat the mixture to boiling, and add to it two or three drops of strong nitric acid to convert the iron into sesquioxide. Then add, little by little, ammonia-water to the boiling liquor until a distinct odor of ammonia is perceptible after the mixture has been thoroughly stirred.

29. *Analysis of the Mixed Precipitate.* The following method of detecting iron, chromium, aluminum (and manganese) in the mixed precipitate which may contain all of them, together with phosphates and other compounds of barium, strontium, calcium and magnesium, depends :—1st, Upon the sparing solubility of the oxalates of barium, strontium, calcium and magnesium in water acidulated with oxalic acid, and upon the ready solubility of ferric oxide and the oxides of chromium, aluminum and manganese in that liquid. 2d, Upon the fact that Prussian blue is formed when a solution of ferrocyanide of potassium is added to the solution of a ferric salt. 3d, Upon the conversion of the oxide of manganese, chromium, or aluminum into manganate, chromate, or aluminate of sodium (or potassium), when fused with a mixture of carbonate of sodium and saltpetre ; and the facts that manganate of sodium has a peculiar green, and chromate of sodium a peculiar yellow color. 4th, That chromate of lead is thrown down as a bright yellow powder when the solution of the chromate of an alkali-metal is added to one of acetate of lead. 5th, That hydrate of aluminum may be isolated in the state of a peculiar flocculent precipitate.

The details of the treatment of the precipitate produced by ammonia-water, are as follows :—Collect the precipitate upon a filter, wash it two or three times with water, transfer it with a spatula to a porcelain dish, pour into the dish twice as much of an aqueous solution of oxalic acid (§ 104) as would be sufficient to cover the precipitate, and boil the mixture during four or five minutes, taking care to add, little by little, water enough to replace that lost by evaporation. At last pour into the boiling solution half its own volume of water, allow the mixture to become cold, collect upon a filter the insoluble oxalates (of members of Classes VI and VII) which have been deposited, wash them thoroughly with water, dry the filter and its contents, and preserve them for future examination in connection with Class VI. (See § 39.)

Transfer a small portion of the filtrate to a test-tube, mix with a few drops of dilute sulphuric acid, and add a drop of a solution of ferrocyanide of potassium. The liquid will immediately become colored with Prussian blue.

Evaporate the rest of the filtrate to dryness in a porcelain dish, taking care to stir the liquid constantly, so that none of the material shall be thrown out of the dish and lost by tumultuous boiling, and ignite the residue to destroy the oxalic acid. Take out a few grains of the residue, boil them in a little strong chlorhydric acid, then dilute with water, and add a drop or two of ferrocyanide of potassium to the solution. A blue precipitate falls, which is a somewhat more emphatic proof of the presence of iron than the blue coloration above described. In case much iron be present, the blue color may be too deep to be recognized until a large quantity of water has been poured into the tube. When the dish has become cold, cover the residue with two or three times its bulk of a dry mixture of equal parts of carbonate of sodium and nitrate of potassium (§ 128). Rub the materials together strongly with the finger, transfer the mixture to a piece of platinum foil, and fuse it *thoroughly* in a strong oxidizing blow-pipe flame (§ 167). In case the quantity of material be large, it may be fused in a platinum crucible at the blast-lamp, or upon the foil, as before, in several successive portions.

Test for **Fe.**

If only a manganese compound, and no chromium, had been fused upon the foil, the cold mass would have exhibited the peculiar bluish-green color of manganate of sodium. If only chromium had been present, the bright yellow color of chromate of sodium would have been clearly perceived. But from mixtures of the manganate and chromate of sodium, in various proportions, different shades of green, brown-green, or yellow-green, will result. When iron is present, the red color of its oxide may obscure the colors due to manganese and chromium.

Place the platinum foil in a porcelain dish, cover it with water, and boil the latter until all the soluble matter has been dissolved from the foil. Take out the clean foil, and throw the contents of the dish upon a filter. The insoluble oxides of iron and manganese remain upon the filter, while the chromate and aluminate of sodium pass into the filtrate, which is colored yellow by the chromium salt.

Wash the insoluble residue with water, allow it to drain, and then fuse a small quantity of it with twice its bulk of a mixture of carbonate of sodium and nitrate of potassium upon platinum foil in the oxidizing blow-pipe flame. The peculiar bluish-green coloration of manganate of sodium will appear in the fused mass, as soon as it has become cold, particularly at the edges and thinner portions. In thus testing for manganese, it is well to incline the foil, so that portions of the thoroughly melted mass may flow away from the centre of the mixture into thin sheets, in order that the color of the manganate may be exhibited in its purity.

Test for **Mn.**

Return now to the filtrate from this insoluble residue. Divide it into two portions. Carefully add acetic acid, drop by drop, to one portion of the filtrate until the liquor exhibits an acid reaction, and then add to it two or three drops of a solution of acetate of lead (§ 137). An insoluble precipitate of chromate of lead will be immediately thrown down, exhibiting a bright yellow color if the reagents be all pure. But if, as often happens, the carbonate of sodium, employed as a flux, was contaminated with sulphate of sodium, the yellow color of the precipitate will tend towards white, in proportion to the amount of sulphate of

Test for **Cr.**

lead which has gone down together with the chromate. A pure white precipitate would be no indication of chromium, but only of a sulphate in the reagents.

Acidulate the other portion of the aqueous solution of the fused sodium compounds with dilute chlorhydric acid, add ammonia-water to slight alkaline reaction, warm the mixture and leave it at rest for at least half an hour, or, better, over night. After the lapse of some time, a characteristic, gelatinous, colorless agglomeration of particles of hydrate of aluminum will appear at the top or bottom of the liquid. It should be said, that flocks of hydrate of aluminum, when diffused through a liquid, are almost transparent enough to elude observation.

When an acid solution, containing much aluminum, is mixed with ammonia-water and warmed, a copious precipitate of hydrate of aluminum will appear immediately, and will often remain floating for some time upon the surface of the solution by virtue of bubbles of air entangled in it. But since it is not easy to convert the whole of the alumina in the original precipitate into soluble aluminate of sodium, by fusion with carbonate of sodium in the method above described, the quantity of the hydrate to be thrown down at the final test is often very small, and considerable time must be allowed, in order that every particle of it may separate from the solution, and all the particles collect into a single mass.

To confirm the presence of aluminum, collect the hydrate in the point of a small filter and allow it to drain. Cut away the superfluous paper, place that portion of the filter to which the precipitate is attached upon a piece of charcoal, and heat it intensely in the blow-pipe flame. Moisten the residue with a drop of a solution of nitrate of cobalt, and again ignite it strongly. The unfused compound of aluminum, cobalt

Test for Al.

and oxygen, left upon the coal, will exhibit a deep sky-blue color, when allowed to cool. This reaction is useful in distinguishing the hydrate of aluminum from that of glucinum, an element somewhat similar to aluminum though far less abundant. Hydrate of glucinum when ignited with nitrate of cobalt does not yield a pure blue compound, but only a gray mass.

30. An outline of the foregoing operations may be tabulated as follows :—

<table>
<tr><td colspan="4">The General Reagent ([NH_4]HO, mixed with NH_4Cl) of Class IV precipitates the hydrates of Fe, Cr, and Al, together with Mn (as a chromite, ferrite or aluminate), and various phosphates and other compounds of Fe, Cr, Al, Ba, Sr, Ca, and Mg. Boil the precipitate in a solution of oxalic acid :—</td></tr>
<tr><td rowspan="3">Oxalates of Ba, Sr, Ca, and Mg separate in the form of a powder, which is collected for future examination in connection with Class VI.</td><td colspan="3">Oxalates of Fe, Al, Cr, and Mn go into solution. Test a small portion of the solution for Fe with ferrocyanide of potassium. Evaporate the remainder to dryness and ignite. Test again for Fe. Mix the residue with dry Na_2CO_3 and KNO_3, fuse upon platinum foil, and treat with boiling water :—</td></tr>
<tr><td rowspan="2">Test the insoluble residue for Mn.</td><td colspan="2">Divide the solution into two portions :—</td></tr>
<tr><td>Yellow color of aqueous solution indicates Cr. Confirm Cr by precipitation of $PbCrO_4$.</td><td>Acidulate a portion of the solution with HCl, and add (NH_4)HO :— Colorless flocculent precipitate proves presence of Al.</td></tr>
</table>

31. *Separation of Class IV from Class III.* The methods of eliminating Classes I, II, and III from mixtures which contain members of these classes as well as of Class IV, have already been described in §§ 6 and 8.

It is essential to the success of the operation that all

the sulphuretted hydrogen in the filtrate from Classes II and III be expelled, and that any iron which may be contained in the solution be converted to the state of a ferric salt, before the general reagent (NH_4HO) of Class IV is added to the liquid. For sulphuretted hydrogen precipitates all the members of Classes IV and V from alkaline solutions, and the filtrate now in question is, of course, made alkaline, when ammonia-water is added to it. The iron must be oxidized, because ferrous hydrate is somewhat soluble in ammonium-salts, and could not, therefore, be precipitated completely by ammonia-water from the acid filtrate from Classes II and III.

No matter what the condition of the iron may have been in the original solution, it is reduced to the state of ferrous salt by sulphuretted hydrogen. The filtrate from the precipitate produced by sulphuretted hydrogen (the general reagent of Classes II and III) should, therefore, be placed in a porcelain dish, and boiled, until the steam from it ceases to blacken lead paper (§ 138). After the sulphuretted hydrogen has been expelled, three or four drops of strong nitric acid must be added to the liquid, and the mixture boiled for a moment longer to oxidize the iron. If much iron be present, the liquid will turn yellow.

To determine whether the iron in the substance subjected to analysis was originally in the state of a ferric or a ferrous salt, test a small quantity of the original solution with a drop of ferricyanide of potassium (§ 126). The formation of Prussian blue proves the presence of a ferrous salt. Another small portion of the original solution, tested with a drop of ferrocyanide of potassium, would yield Prussian blue in case the solution contained a ferric salt. In applying either of these tests, the blue coloration, indicative of iron, is alone to be looked for ;

no notice need be taken of other colorations, or of precipitates formed by the action of the ferri- or ferro-cyanide upon the various metallic salts which the solution may contain. The possibility that a ferrous salt may have been changed into a ferric salt during the process of getting the original substance, if a solid, into solution, must not be lost sight of.

After the iron has been oxidized, a small quantity of a solution of chloride of ammonium is added to the boiling liquid, and finally ammonia-water, little by little, with constant stirring, until a persistent odor of ammonia is perceptible. A large excess of ammonia must be carefully avoided, for hydrate of aluminum, being somewhat soluble in ammonia-water, might be kept in solution, to the disturbance of the analysis of Classes VI and VII.

It will be remembered that the object of using chloride of ammonium is to hold in solution magnesium (of Class VII) and the members of Class V. A considerable quantity of the ammonium-salt will, of course, be formed in any event by the action of the ammonia-water upon the chlorhydric acid in the solution, but it is best always to add a further portion of the chloride, as a precautionary measure.

The following inferences may be drawn from the color of the precipitate produced by ammonia-water :—

A gelatinous white precipitate indicates aluminum or phosphate of calcium.

A grayish-green or grayish-blue precipitate indicates chromium.

A reddish-brown precipitate indicates iron.

If no precipitate is produced by the ammonia-water, all the members of Class IV are absent, and the solution may at once be tested with sulphydrate of ammonium, the general reagent of Class V.

When the solution contains much chromium, a small portion of this element is apt to remain dissolved at first in the excess of ammonia-water, and to color the solution pink ; but by continuing to boil the solution, the color may be made to disappear, and the whole of the chromium thrown down. Care must be taken to replace, by small portions, the water driven off by boiling, lest some of the members of Class V be rendered insoluble.

It is to be observed that the legitimate members of Class IV cannot be completely precipitated by ammonia-water from solutions which contain non-volatile organic substances, like albumen, sugar, starch, and so forth, or organic acids (such as tartaric, citric, oxalic, or even in some cases acetic acid) which form soluble double salts by uniting simultaneously with the ammonium and one or more of the members of the class. The treatment of substances containing organic matter will be explained hereafter. (§ 76, I.)

CHAPTER VI.

CLASS V. SULPHIDES INSOLUBLE IN WATER AND IN SALINE OR ALKALINE SOLUTIONS.

32. *Example of the Precipitation of the Members of Class V.* Place in a small glass flask six or eight drops of strong aqueous solutions of the sulphates, nitrates, or chlorides of cobalt, nickel, manganese and zinc. Add to the mixture five or six teaspoonfuls of a solution of chloride of ammonium, twice as much water, and ammonia-water to alkaline reaction.

Heat the mixture to boiling, and add sulphydrate of

ammonium to the boiling solution, drop by drop, with frequent agitation, as long as a precipitate continues to be formed. (Compare p. 33.) In the present case there are special reasons why the precipitate should be boiled and shaken, in order to make it compact; for the sulphides of Class V, when loose and flocculent, are not only easily acted upon by the air and by dilute acids, but are peculiarly liable to pass through the pores of filter-paper and yield muddy filtrates.

At the best, these sulphides oxidize rapidly when moist, with formation of soluble sulphates which are liable to pass through the filters and contaminate the filtrates. The analysis of the sulphides should therefore be proceeded with immediately after the precipitation with sulphydrate of ammonium, and should be conducted in such manner that no precipitate of a sulphide shall ever be left moist upon a filter more than half an hour.

33. *Analysis of the Mixed Sulphides.* The detection of the several members of Class V depends, 1st, Upon the almost complete insolubility of the sulphides of cobalt and nickel in cold dilute chlorhydric acid, and the ready solubility of the sulphides of manganese and zinc in that liquid. 2d, Upon the solubility of hydrate of zinc, and the insolubility of hydrate of manganese, in a solution of caustic soda. 3d, Upon the insolubility of sulphide of zinc in alkaline solutions. 4th, Upon the peculiar colors imparted to borax glass by compounds of cobalt and nickel dissolved in the glass; and upon certain other special tests to be described directly.

To effect the separation:—Collect the precipitate upon a filter, and rinse it once or twice with water; spread open the filter in a porcelain dish, and cover it with cold dilute chlorhydric acid. Scarcely any of the sulphide of cobalt, or of nickel, will go into solution, while the sulphides of manganese and zinc will be completely decomposed, and dissolved as chlorides.

Filter the chlorhydric acid solution, pour the filtrate into a porcelain dish, and boil it, until strips of moistened lead paper held in the steam no longer indicate the presence of sulphuretted hydrogen ; then transfer the liquid to a beaker or test-tube, and add caustic soda (§ 116) to it in slight excess. A whitish gelatinous precipitate of hydrate of manganese, insoluble in caustic soda, will be thrown down, together with small portions of the hydrates of cobalt and nickel (particularly the latter), resulting from the partial decomposition of the sulphides of these metals by the chlorhydric acid, while the hydrate of zinc at first precipitated redissolves completely in the excess of soda. It is to be observed that precipitation should never be effected in a porcelain dish, since a white or transparent precipitate is scarcely visible in a white and opaque dish.

To prove the presence of manganese, collect the precipitate upon a filter, allow it to drain, and fuse a small portion of it with a mixture of carbonate of sodium and nitrate of potassium upon platinum foil in the oxidizing blow-pipe flame, as directed on p. 49.

Through the alkaline filtrate pass sulphuretted hydrogen gas. Sulphide of zinc will be thrown down as white or dirty-white flocculent precipitate.

To confirm the presence of zinc :—Collect the precipitate produced by sulphuretted hydrogen upon a filter and allow to drain ; transfer it to a porcelain dish, dissolve it in dilute chlorhydric acid, and evaporate the solution almost to dryness on a water-bath (§ 166). Dissolve the residue in a few drops of water, and without heeding the milkiness which the presence of particles of free sulphur may produce in it, pour this liquid into a test-tube containing two or three teaspoonfuls of a solution of normal chromate of potassium previously heated to boiling. A peculiar yellow, somewhat flocculent precipitate of basic chromate of zinc will be formed in the boiling liquid, and will soon subside to the bottom of the tube. The red color of the supernatant fluid is due to the formation of bichromate of potassium.

Test for **Zn.**

It is to be observed that in the analysis of mixtures which contain no manganese, the precipitate of hydrate of cobalt or of nickel, produced by the caustic soda, is usually small and sometimes hardly perceptible ; but, no matter how minute the precipitate may be, it must always be carefully removed by filtration before testing the solution for zinc with sulphuretted hydrogen.

The black residue, insoluble in dilute chlorhydric acid, is washed with water and tested for cobalt and nickel, by heating successive small portions of it in a bead (§ 84, *c.*) of borax (§ 119) in the oxidizing blow-pipe flame. If cobalt alone were present, a bright, pure blue color would be imparted to the bead. On the other hand, if the precipitate was composed solely of sulphide of nickel, the borax glass would assume a peculiar reddish-brown color. Mixtures of the two sulphides yield beads of various tints, according to the proportions of nickel and cobalt contained in them. By adding the precipitate to the borax by repeated small portions, and fusing the bead anew after each addition, it is often possible to obtain first the characteristic color of one of the elements and afterwards tolerably well defined indications of the other.

Tests for **Co & Ni.**

The blue color of cobalt can usually be made manifest, even in presence of much nickel, by heating the borax bead in the reducing blow-pipe flame (§ 167). In the reducing flame the reddish-brown color imparted by nickel changes to gray, while the cobalt blue remains unaltered.

In any event, one of the two metals will be detected by the blow-pipe test, and the subsequent operations can be limited to searching for the other.

To prove the presence of nickel, boil the black residue with a few drops of aqua regia in the porcelain dish, and evaporate the solution almost, but not quite, to dryness. Add to the residual acid liquor, little by little, a strong solution of cyanide of potassium, until the reaction of the solution becomes decidedly alkaline, and boil the mixture for five minutes, taking care to add water by small portions to fully replace that lost by evaporation. The cyanides of nickel and cobalt, at first thrown down, both redissolve easily in an excess of cyanide of potassium; but while the cyanide of nickel undergoes no change when the mixture is boiled, the cyanide of cobalt is all converted into cobalticyanide of potassium, and from solutions of this compound, cyanide of cobalt is not precipitated on the addition of acids.

As soon as the liquid has become cold, pour dilute sulphuric acid into it, without heeding any precipitate which may have formed during the evaporation, until a drop of the mixture turns blue litmus-paper red. Transfer the acidulated liquor to a large test-tube, fill the latter with water, shake the mixture well, and allow it to stand during eighteen or twenty-four hours. The soluble compound of cyanide of nickel and cyanide of potassium is decomposed by the sulphuric acid, and the cyanide of nickel precipitated in the form of a light, dirty greenish-yellow powder, which slowly subsides to the bottom of the tube.

Test for Ni.

Sometimes a dark layer of dirt, derived from impurities in the reagents, is deposited above or below the stratum of cyanide of nickel, but it seldom happens that the characteristic color of the latter is materially obscured by this contamination. At other times, when the operations have been performed carelessly, and the reagents have been employed in undue quantities, crystals of sulphate of potassium will separate in the tube ; but they can readily be removed by dissolving them in water.

To confirm the presence of cobalt in case of doubt :— Dissolve the black residue in a few drops of hot aqua regia, evaporate the solution nearly to dryness, pour into the residual solution two or three times its own volume of a solution of nitrite of potassium (§ 129), and add to the mixture concentrated acetic acid, until the reaction of the liquid is strongly acid. Transfer the mixture to a test-tube, and leave it at rest during eighteen or twenty-four hours. A beautiful, yellow crystalline precipitate of the double nitrite of cobalt and potassium will be deposited sooner or later, according to the proportion of cobalt which the solution contained.

Test for Co.

On adding caustic soda to the filtrate from the yellow cobalt precipitate, hydrate of nickel would be thrown down if it were present, and the presence of nickel might be confirmed by testing this precipitate with borax in the oxidizing blow-pipe flame.

34. An outline of the foregoing operations may be tabulated as follows :—

<table>
<tr><td colspan="3">The General Reagent ($[NH_4]HS$) of Class V precipitates CoS, NiS, MnS, and ZnS. Treat the precipitate with dilute HCl :—</td></tr>
<tr><td rowspan="2">CoS and NiS remain undissolved.
Test for Co and Ni with borax glass, and, if need be, with KCN or KNO_2.</td><td colspan="2">$MnCl_2$ and $ZnCl_2$ go into solution. Boil, to expel H_2S, and add NaHO :—</td></tr>
<tr><td>Hydrate of manganese is precipitated, together with traces of the hydrates of Co and Ni.
Prove presence of Mn by the blow-pipe test.</td><td>Hydrate of zinc goes into solution. Add H_2S to throw down ZnS.
Confirm presence of zinc by precipitation of the chromate.</td></tr>
</table>

35. *Separation of Class V from Class IV.* After Classes I, II, III, and IV have been removed in the manner already described (§§ 9, 31), add a single drop of sulphydrate of ammonium of good quality (§ 110) to the filtrate from Class IV. If no precipitate is produced, none of the members of Class V can be present, and the solution may be immediately tested with carbonate of ammonium, the general reagent of Class VI.

If the first drop of the sulphydrate produces a precipitate, transfer the mixture to a small flask, heat it until it actually boils, and add more of the sulphydrate, with the precautions enjoined on p. 55 to complete the precipitation.

In case the precipitate produced by sulphydrate of ammonium be white, the presence of zinc is indicated.

If it be flesh-colored or yellowish-white, and become brown by oxidation when exposed to the air, the presence of manganese is to be inferred.

In case the precipitate is black, either cobalt or nickel, or both these elements, are present. Both of them must be sought for, whenever the precipitate exhibits any tinge of black at the moment of its formation.

CHAPTER VII.

CLASS VI. CARBONATES INSOLUBLE IN WATER, AMMONIA-WATER, AND SALINE SOLUTIONS.

36. *Example of the Precipitation of the Members of Class VI.* Place in a test-tube six or eight drops of strong aqueous solutions of the chlorides or nitrates of barium, strontium, and calcium. Add to the mixture two or three teaspoonfuls of a solution of chloride of ammonium, enough ammonia-water to produce an alkaline reaction, and finally a solution of carbonate of ammonium, drop by drop, as long as any precipitate continues to be produced by fresh portions of this reagent. To determine this last point, heat the mixture to boiling at intervals, and after boiling allow it to settle, until a sufficient quantity of comparatively clear liquid has collected at the top of the mixture to permit the application of the test.

37. *Analysis of the Mixed Carbonates.* The separation of barium, strontium and calcium, one from the other, depends :—1st, Upon the insolubility of chromate of barium in dilute acetic acid, and the solubility of the chromates of strontium and calcium in that liquid. 2nd, Upon the fact that sulphate of strontium is almost absolutely insoluble in acidulated water, while sulphate of calcium, though rather sparingly soluble in water, is still sufficiently soluble to be kept in solution. (See § 123.)

Collect the precipitate upon a filter, wash it two or three times with water, taking care to collect the precipitate at the apex of the filter, and dissolve it in acetic acid. The acid may be poured into the filter as it rests in the funnel, but only a moderate quantity should be used, and the filtrate should be poured back upon the filter, until the acid is saturated. Finally rinse the filter with a little water from a wash-bottle with small orifice, collect the wash-water with the filtrate, and shake the mixture.

Pour a small portion of the acetic acid solution into a test-tube, and add to it a drop of a solution of normal chromate of potassium. A pale yellow precipitate falls when barium is present, as in this instance; for chromate of barium is well nigh insoluble in acetic acid, especially in presence of saline solutions. In order to separate the whole of the barium, pour the contents of the test-tube into the reserved portion of the acetic acid solution, heat the mixture to boiling, and add to it chromate of potassium, until no more precipitate falls and the supernatant liquor appears distinctly colored, after having been well shaken and allowed to settle. Filter the mixture, and proceed to examine the filtrate for strontium and calcium.

Test for **Ba.**

If no barium had been present, no precipitate would have been produced by chromate of potassium in the small portion of liquid first tested, and it would have been unnecessary to mix this reagent with the rest of the acetic acid solution. Trouble would thus be saved, as will appear below.

It sometimes happens that chromate of barium is precipitated in the form of powder so fine that some particles of it pass through the pores of the paper and contaminate the filtrate. Now, in order to detect strontium and calcium, it is absolutely necessary that this filtrate, although of a bright yellow color, should be perfectly transparent and free from suspended particles of the barium salt. If then the filtrate is at all turbid, it must be poured back repeatedly into the filter, and again collected in clean tubes, until the last trace of cloudiness has disappeared.

To the filtrate from the chromate of barium add ammonia-water to alkaline reaction, and carbonate of ammonium as long as a precipitate falls. Heat the mixture to boiling for a moment, collect the precipitate upon a small filter, and wash it with water, until all the chromate of potassium has been removed, and the wash-water runs colorless from the filter.

Tests for **Sr. & Ca.**

Dissolve the precipitate in the smallest possible quantity of acetic acid, and mix it with three or four times its volume of a solution of sulphate of potassium (§ 123), made of such strength that, though capable of throwing down sulphate of strontium, it cannot precipitate sulphate of calcium. Allow the mixture to stand at rest for two hours or more, in order that the white powder of sulphate of strontium may separate completely. Then filter, and to the filtrate add ammonia-water to alkaline reaction, and half a teaspoonful of a solution of oxalate of ammonium (§ 113). A white precipitate of oxalate of calcium will be immediately thrown down.

Since sulphate of strontium is somewhat soluble in a solution of chromate of potassium, the filtrate from chromate of barium cannot be examined directly for strontium by means of sulphate of potassium. The strontium and calcium are consequently reprecipitated as carbonates, in order that the excess of chromate of potassium may be washed away. The operation serves also to collect the strontium and calcium out of the mass of liquid in which they have become diffused, and to concentrate them to a small bulk.

It should be observed that, when the proportion of

strontium or calcium in a mixture is small, it often happens that the precipitate, produced by carbonate of ammonium in the filtrate from chromate of barium, is held in suspension, and concealed so completely in the yellow liquor, that an unpractised eye can hardly detect the fact that the liquid has become cloudy. That a precipitate has really been formed in such cases is easily discovered by throwing a portion of the mixture upon a filter, and comparing the clear filtrate thus obtained with that portion of the mixture which has been left unfiltered.

38. An outline of the foregoing operations may be presented in tabular form, as follows :—

<table>
<tr><td colspan="3">The General Reagent ($[NH_4]_2CO_3$) of Class VI, precipitates the carbonates of Ba, Sr, and Ca. Dissolve in dilute acetic acid, and add K_2CrO_4 :—</td></tr>
<tr><td rowspan="2">$BaCrO_4$ is thrown down as a yellow powder.</td><td colspan="2">$SrCrO_4$ and $CaCrO_4$ remain in solution. Add $(NH_4)_2CO_3$ and $(NH_4)HO$. Collect and wash the precipitate, and dissolve it in acetic acid. Add dilute K_2SO_4 :—</td></tr>
<tr><td>$SrSO_4$ is thrown down.</td><td>$CaSO_4$ remains in solution. Add oxalate of ammonium, to precipitate oxalate of calcium.</td></tr>
</table>

39. *Separation of Class VI from the Preceding Classes.* After Classes I, II, III, IV and V have been separated in the manner already described (§§ 6 to 10), there will still always remain to be examined the filtrate from Class V, and sometimes a precipitate (§ 29) composed of oxalates of barium, strontium, calcium and magnesium,—in case any salt of these elements, insoluble in ammonia-water, has been thrown down with the members of Class IV.

If such a precipitate has been obtained in the analysis of Class IV, the oxalic acid contained in it must now be destroyed, the remainder of the precipitate brought into solution, and this solution added to the filtrate from Class V, before proceeding to precipitate the members of Class VI. To this end, ignite the dry precipitate carefully upon platinum foil—by several successive portions if the precipitate is large,—taking care that none of the powder is left sticking to the paper or lost by dropping it from the foil. At a moderate heat the oxalates suffer decomposition, and only carbonates or oxides are left upon the foil. Place the foil and the residue in a small porcelain dish, and dissolve the residue in boiling dilute chlorhydric acid. Add a few drops of chloride of ammonium to the solution, neutralize the acid with ammonia-water, pour the liquid upon a small filter and add the filtrate to that obtained from Class V. Then add a solution of carbonate of ammonium to the mixture, and boil it in the manner described in § 36.

If there be no precipitate of the oxalates from Class IV, the filtrate from Class V will, of course, be treated directly with carbonate of ammonium, care being taken to add only a drop or two of the reagent, at first, to ascertain whether any of the members of Class VI are really contained in the solution.

The solution to which the general reagent carbonate of ammonium is added, must contain chloride of ammonium, to prevent the precipitation of magnesium as a carbonate, and also ammonia-water, to hinder the decomposition of the carbonates of barium, strontium and calcium, by the boiling chloride of ammonium. But since the excess of ammonia-water and the chloride of ammonium, added to the solution before the separation of Class IV, are still contained in it, no new quantity of either of them need here be added.

It is to be remembered, in this connection, that the carbonates of barium, strontium and calcium, are all slightly soluble in a solution of chloride of ammonium, and that no precipitate whatever is produced when carbonate of ammonium is added to a weak solution of either of the members of Class VI, in case a large quantity of chloride of ammonium has been previously mixed with it.

In a solution containing traces of barium or strontium these elements might fail to be detected, in case the chlorhydric acid employed in the process of separating Classes I and II was contaminated with sulphuric acid, or in case the original liquid contained nitric acid to oxidize a portion of the sulphur of the sulphuretted hydrogen employed to precipitate Class II, or even if the nitric acid, employed to oxidize iron in the filtrate from Classes II and III, were added before the sulphuretted hydrogen had been expelled. All danger could be avoided, however, by using pure chlorhydric acid to precipitate Class I, and expelling the nitric acid from the filtrate by evaporating the latter to dryness upon a water-bath, covering the residue with pure concentrated chlorhydric acid, again evaporating until the mixture became dry, and finally dissolving in acidulated water.

CHAPTER VIII.

CLASS VII.—INCLUDES THE REMAINING COMMON ELEMENTS NOT COMPRISED IN THE PRECEDING CLASSES, NAMELY: MAGNESIUM, SODIUM AND POTASSIUM.

40. *The detection of the Several Members of Class VII* depends :—1st, Upon the insolubility of a double phosphate of magnesium and ammonium, and the solubility of the phosphates of potassium and of sodium; and 2d, Upon the facts that compounds of sodium and potassium impart peculiar colorations to non-luminous flames, like those of alcohol and of a mixture of coal-gas and air.

Prepare a mixture of ten or twelve drops of strong solutions of almost any one of the salts of magnesium, sodium and potassium, and add to the mixture an equal bulk of chloride of ammonium. Pour a quarter of the mixture into a test-tube and the remainder into a small porcelain dish. Add to the contents of the test-tube two or three drops of a solution of diphosphate of sodium (§ 121), and as much ammonia-water, and shake the cold mixture at frequent intervals during one or two hours. A crystalline, white precipitate of phosphate of magnesium and ammonium will appear after a longer or shorter interval, according as the original solution was more or less dilute.

Test for Mg.

41. Evaporate the contents of the porcelain dish to dryness, ignite the residue until the chloride of ammonium has been completely expelled—a point which will be indicated by the cessation of fuming—allow the dish to cool, and pour into it three or four drops of water.

Carefully clean the loop on a piece of platinum wire by washing it repeatedly with water, and finally holding it in the lamp flame until the last traces of sodium compounds are burned off, and it ceases to color the flame. Without touching the loop with the fingers, dip it into the aqueous solution in the dish, and again hold it in the flame. A bright yellow color will be imparted to the flame by the sodium contained in the mixture; but the color peculiar to potassium compounds will be invisible, since the yellow color of the sodium overpowers and conceals it. Dip the loop a second time in the solution, and again hold it in the lamp flame; but this time look at the flame through a piece of deep-blue cobalt glass. This cobalt glass is the ordinary blue glass used for stained glass windows; it is essential that the glass should be of moderate thickness, and colored blue throughout, not simply "flashed" with blue. The characteristic violet color imparted to flame by potassium compounds will now be visible, for the blue glass shuts off completely the yellow sodium light, while it permits the free passage of the violet rays.

Test for Na.

Test for K.

Since traces of compounds of sodium and potassium are to be found almost everywhere, it is sometimes difficult to determine by the foregoing tests whether the substance under examination contains one or other of these elements as an essential ingredient, or merely as an accidental impurity. It is always possible, however, to separate the sodium or the potassium from the other members of the class, and to decide, by actual inspection of the isolated compounds, whether one or both of these substances is contained in really appreciable quantity in the substance subjected to analysis. If only potassium is sought for, filter the aqueous solution, last mentioned, through a very small filter, add to the filtrate a drop or two of chlorhydric acid and several drops of a solution of bichloride of platinum (§ 146). A yellow crystalline precipitate of chloroplatinate of potassium will separate after some time. Since chloride of ammonium would produce a similar precipitate, it is, of course, essential that the ammoniacal salt should be expelled by ignition before potassium can be tested for.

In case sodium, or both sodium and potassium, be sought for, the magnesium must first be got rid of. To this end, moisten the dry residue, which contains the chlorides of magnesium, of potassium and of sodium, with a drop or two of water, mix it thoroughly with an equal bulk of red oxide of mercury, and ignite the mixture until fuming ceases. The chloride of magnesium will be changed to oxide, while the easily volatile chloride of mercury escapes :—

$$MgCl_2+HgO=MgO+HgCl_2.$$

The ignition should be effected beneath a chimney, or in a draught of air powerful enough to carry away the poisonous fumes of the corrosive sublimate. Boil the ignited residue with a small quantity of water ; separate the insoluble oxide of magnesium, together with the excess of oxide of mercury, by filtration ; evaporate the filtrate to a small bulk ; throw down the potassium as chloroplatinate, collect the latter upon a filter, and from the filtrate remove the excess of chloride of platinum by means of sulphuretted hydrogen. Filter off the sulphide of platinum, and evaporate the filtrate to dryness to obtain the chloride of sodium. Or, instead of employing sulphuretted hydrogen, evaporate the filtrate from chloroplatinate of potassium in a watch glass, at a gentle heat, until the liquid begins to become dry at its edges, and then examine it with a magnifying glass. Characteristic crystals of chloroplatinate of sodium will be seen in the form of long, slender prisms or needles of yellow color.

42. *The Isolation of Class VII,* by the removal of the preceding classes, has been described in § 12. Care must always be taken to concentrate the whole of the filtrate from Class VI by evaporation, before testing a portion of it for magnesium ; and time enough must be allowed for the magnesium precipitate to crystallize. The remainder of the filtrate from Class VI must be evaporated to dryness and ignited, before testing for sodium and potassium, in order that the flame reactions of these elements may not be concealed or obscured by the vapors of ammonium-salts, or the combustion of particles of organic matter derived from the various reagents which have been added in the course of the analysis.

43. An outline of the methods employed for separating the several classes is presented in tabular form on the opposite page. The method of separating Class III from Class II must be slightly modified when mercury is present in the form of a mercuric salt. Sulphydrate of ammonium should in that case be substituted for sulphydrate of sodium ; because the sulphide of mercury is rather soluble in sulphydrate of sodium. The presence of mercury will generally be made known during the preliminary examination (§ 76, III, c.). Sulphide of copper also is somewhat soluble in the sulphydrates of sodium and ammonium, in presence of the sulphides of Class III ; but enough of this sulphide will always remain undissolved to insure the detection of copper in Class II.

Since sulphydrate of ammonium often fails to dissolve

TABLE FOR THE SEPARATION OF THE SEVEN CLASSES OF THE METALLIC ELEMENTS.

<table>
<tr><td colspan="7">Add an excess of dilute HCl to the solution to be examined.</td></tr>
<tr><td rowspan="5">A precipitate indicates

Pb
Ag
Hg

(See § 17.)</td><td colspan="6">Saturate the filtrate with H_2S.</td></tr>
<tr><td colspan="2">Boil the precipitate with NaHS.</td><td colspan="4">Boil the filtrate, to expel H_2S, and add HNO_3, to oxidize Fe. Then add NH_4Cl and a slight excess of NH_4HO.</td></tr>
<tr><td rowspan="3">A residue indicates

Hg
Pb
Bi
Cd
Cu

(See § 20.)</td><td rowspan="3">To the alkaline filtrate add an excess of HCl.
——
A precipitate indicates

As
Sb
Sn
[Au
Pt]

(See §§ 24, 43.)</td><td rowspan="3">A precipitate indicates

Fe
Cr
Al
3CaO, P_2O_5
etc.

(See § 29.)</td><td colspan="3">To the filtrate add a drop of NH_4HS.</td></tr>
<tr><td rowspan="2">A precipitate indicates

Co
Ni
Mn
Zn

(See § 33.)</td><td colspan="2">To the filtrate add a drop of $(NH_4)_2CO_3$.</td></tr>
<tr><td>A precipitate indicates

Ba
Sr
Ca

(See § 37.)</td><td>Evaporate the residual liquor to a small bulk, and test a portion of it for Mg (see § 40).
Evaporate the rest of the solution to dryness; ignite and test for Na and K (see § 41).</td></tr>
</table>

sulphide of tin, it is not, in general, so fit a solvent for the sulphides of Class III as sulphydrate of sodium.

CHAPTER IX.

GENERAL TESTS FOR THE NON-METALLIC ELEMENTS.

44. The following common elements remain to be considered :—Sulphur, nitrogen, phosphorus, carbon, boron, silicon, chlorine, bromine, iodine, fluorine, oxygen, hydrogen. It is obvious that oxygen and hydrogen cannot be directly sought for by any analytical process conducted in the wet way. These elements are added to the original matter in the water which is used as a solvent. The presence of these elements is either inferred from the other results of the analysis, or forms the object of direct inquiry in the *preliminary treatment*—that important part of every analysis which forms the main subject of Part II. The remaining elements all belong to that class vaguely described as *non-metallic ;* they unite with oxygen, hydrogen, or both these elements, to form what are commonly called *acids.*

As the metallic elements are recognized through familiar compounds, oxides, chlorides, sulphides, or other, so these non-metallic elements are identified by working out of the unknown mixture their characteristic compounds. These compounds are generally salts. But there is one marked difference between the search for the metallic and the search for the non-metallic elements in the wet way. In connection with the determination of a metallic element, no question usually arises except the fundamental one of its presence or absence in a given solution. The sodium in the three different salts, sulphate, sulphite and hyposulphite of sodium, for example, is detected in one and the same method ; but, when the other elements of these salts are sought for, three different reactions will occur, according to the varying nature of the ingredients other than sodium. A sulphate in solution gives one set of reactions, a sulphite another, and a hyposulphite a third. It is important to do more than determine the mere presence of sulphur and oxygen. We want to know whether the sodium salt be a sulphate, sulphite, or hyposulphite. Analogous questions arise with regard to other non-metallic elements ; there is chlorine both in chlorides and chlorates, and carbon both in carbonates and oxalates. Arsenic, too, may be present in the form of arsenite or arseniate, and these two different kinds of salts exhibit many quite dissimilar reactions.

The examination into the nature of the other ingredients of any compound is invariably subsequent to the determination of the metallic element or elements by the method detailed in the preceding chapters. In the language of the dualistic theory, when the *base* has been found, the *acid* is sought for. By whatever words the object of the analyst be described, the process itself is in general as follows :—The analyst tries to identify each class of salts by precipitating, or otherwise making manifest, some familiar member of that class. Thus he identifies sulphates by precipitating sulphate of barium, chlorides by precipitating chloride of silver, carbonates by throwing down carbonate of calcium, and so forth. The practically important inquiry is, how to find means of identifying each of the principal classes or kinds of salts. The word " class " or " kind " of salts is used in this connection as a collective name for all the salts from which the same acid may be derived, or which, in dualistic language, "contain" the same acid: thus all sulphates constitute one class or kind, all carbonates another, and so forth. The following common classes of salts are those for which more or less perfect means of recognition will be given in this chapter:—Sulphates, sulphites, hyposulphites, sulphides, arseniates, arsenites, phosphates, carbonates, oxalates, tartrates, borates, silicates, chromates, fluorides, chlorides, bromides, iodides, cyanides, nitrates, chlorates, acetates.

No system of successive testing and elimination, analogous to that already described for the metallic elements, has been devised for the non-metallic constituents of salts. There are, indeed, certain so called *general reagents* for *acids;* but these reagents are chiefly useful to show the simultaneous presence or absence of members of *several* classes of salts, and hardly help towards the identification of any individual class, except in cases of the simplest sort in which only a single class of salts is represented.

The knowledge of the metallic element or elements present, previously gained, is usually a great help in the determination of the other constituents. A single example will illustrate sufficiently for our present purpose this important principle, which is of very wide application in qualitative analysis. Suppose calcium to have been found in an aqueous solution of various salts; it is directly to be inferred with great confidence, though not with entire certainty, for example, that there is no sulphate, phosphate, silicate, carbonate, oxalate, or tartrate in the original liquid, since these calcium salts are insoluble in water; that, in short, the greater number of the classes of salts above enumerated are absent. The list of salts excluded by the demonstrated presence of calcium would be very long. So it is in greater or less degree with many other metallic elements. It is, indeed, no easy matter to make a solution containing even one representative of the seven classes into which the metallic elements have been divided, because each of these elements, except sodium and potassium, present in solution excludes one or more whole classes of salts. By attending to the just inferences to be drawn from the quality of the metallic elements, much time will be saved, and the want of a systematic procedure in searching for the non-metallic elements will be little felt. The presence of the arsenites and arseniates, of chromates, sulphides, sulphites, and hyposulphites will ordinarily be revealed in the course of the search for the metallic element.

Before giving the special tests by which the above-mentioned classes of salts are identified, we proceed to describe certain general tests which are of value, particularly when they give negative results.

45. *The Barium Test.* When a solution of chloride (or nitrate) of barium is added in suitable quantity to a neutral or slightly alkaline solution containing representatives of any or all of the following classes of salts, precipitation usually occurs, for all these salts of barium are insoluble in water and alkaline liquids, if no ammonium-salts be present:

Sulphates,	Arseniates,	Tartrates,
Sulphites,	Arsenites,	Borates,
Hyposulphites,	Carbonates,	Silicates,
Phosphates,	Oxalates,	Chromates,
	Fluorides.	

If the chloride (or nitrate) of barium fail to produce a precipitate under the prescribed conditions, the complete absence of all the above thirteen classes of salts is at once demonstrated, provided that no ammonium-salts be contained in the original solution.

46. *Illustration of the Barium Test.* Prepare in a test-tube a solution containing at once sulphate, phosphate, and carbonate of sodium; a very small bit of each salt will be sufficient. The solution will be found to be alkaline to litmus paper. Add to it chloride of barium (§ 135), little by little, until a fresh addition of the reagent no longer produces an additional precipitate. The white precipitate consists of sulphate, phosphate and carbonate of barium. All the thirteen barium salts which are liable to precipitation under these circumstances are white, with the single exception of chromate of barium. The yellow color of the chromate of barium (§ 36) distinguishes this one precipitate from all the rest. If this color is well marked, the presence of a chromate in the original solution (which must also have been yellow) may be inferred with certainty. In all other cases, however, the precipitate is white, as in the present experiment. The next question is, can anything further be learned from this white precipitate?

Add to the contents of the test-tube dilute chlorhydric acid, until the liquid has a decidedly acid reaction to litmus paper. An effervescence indicates the escape of carbonic acid, displaced by the less volatile chlorhydric acid. The bulky precipitate which the chloride of barium produced will in part disappear, but a portion of it remains undissolved. Filter the contents of the tube. The particles of precipitated sulphate of barium are so very fine that they often pass through the pores of the paper, necessitating repeated filtration through the same filter. To the filtrate add ammonia-water until the liquid has an alkaline reaction. A precipitate will reappear. Of all the barium salts which may be precipitated under these circumstances, only one, the sulphate of barium, is insoluble in chlorhydric and other strong acids. A separation of sulphur from a hyposulphite will not be mistaken for a barium precipitate (§ 22, p. 37). The fact that any of the original precipitate remains undissolved by the chlorhydric acid demonstrates the presence of a sulphate. The portion of the original precipitate which dissolved in the acid, and was reprecipitated by ammonia, consisted in this particular experiment of phosphate of barium; but in an actual analysis the possible salts represented would be so numerous as to make the indication of but little value.

If ammonia-water should cause no reprecipitation in the acid liquid which was filtered from the original precipitate, it must not be inferred that the acid of course dissolved nothing. The borate, oxalate, arseniate, arsenite, tartrate and fluoride of barium are all moderately soluble in solutions of ammoniacal salts, and may not be precipitated on the addition of ammonia. Of course, if the original solution contained ammonium-salts, these six barium salts, if present in small quantity only, might fail to be precipitated. In fact, all the thirteen above-mentioned barium salts, except the sulphate, are more or less soluble in solutions of ammonium-salts, so that whenever these ammonium-salts are known to be present, there is really but one perfectly satisfactory indication with chloride of barium. The presence of a sulphate is revealed by it with certainty, but the results of the other tests must be received with some distrust. If the original solution be acid, it is necessary to neutralize it with ammonia-water before the chloride of barium is added. Sometimes a precipitate is produced by the ammonia-water so added; in that case it is necessary to filter and proceed with the filtrate, although the ammonia-water may have thrown down salts of several of the classes above named, such as phosphates, oxalates, and fluorides. Even if the ammonia-water produce no precipitate, the testing is then performed under the disadvantage of the presence of ammonium-salts.

If the original solution contained silver or lead salts, or mercurous salts, it would be impossible to use *chloride* of barium and chlorhydric acid as reagents ; they would throw down the chlorides of those metals. Nitrate of barium (§ 136) and dilute nitric acid must then be used.

The acids added to the precipitate formed by the barium salt must be always dilute acids. Chloride and nitrate of barium are themselves insoluble in concentrated chlorhydric and nitric acids, and if a strong acid were used as a solvent, the reagent salt might itself separate from the liquid.

47. *The Calcium Test.* Chloride (or nitrate) of calcium precipitates the same classes of salts as chloride (or nitrate) of barium, with the single exception of the chromates. When *sulphates and all ammoniacal salts are absent,* or present only in minute quantities, something may be learned by testing a neutral or slightly alkaline solution supposed to contain representatives of some of the other classes of salts enumerated in § 45, with chloride or nitrate of calcium. The calcium salts liable to precipitation under these circumstances are all soluble in acetic acid, except the oxalate and the fluoride. The precipitate produced by the calcium reagent is, therefore, treated with acetic acid ; if it completely redissolves, *oxalates* and *fluorides* are most probably *absent.* The presence of notable quantities of ammonium-salts renders this test of uncertain value ; because the fluoride and many other salts of calcium are soluble in solutions of ammonium-salts. Since sulphate of calcium is sparingly soluble in water and acetic acid, the presence of a sulphate, causing precipitation of sulphate of calcium, obscures the reaction for oxalates and fluorides. Nitrate of calcium must be used instead of the chloride whenever silver or lead salts, or mercurous salts, are present in the solution under examination.

48. *Illustration of the Calcium Test.* Prepare in a test-tube an aqueous solution of phosphate, oxalate and tartrate of sodium. A very small quantity of each salt will be enough. The solution will be neutral or faintly alkaline. Add to this solution a solution of chloride of calcium (§ 134) until the precipitation is complete. Collect the white precipitate upon a filter, and, when drained, transfer it to a test-tube and treat it with acetic acid. The phosphate and tartrate of calcium will redissolve, but the oxalate remains untouched. To verify this result, and identify each one of the sodium salts present in the original solution, special tests, to be hereafter described, must be resorted to.

49. *The Silver Test.* Nitrate of silver produces a precipitate in neutral or acid solutions with all chlorides, bromides, iodides and cyanides, and in neutral solutions with most of the classes of salts enumerated in § 45. In order to obtain the most comprehensive negative conclusion in the case that nitrate of silver produces no precipitate, it is necessary to operate upon a neutral solution. If the original solution be neutral, the nitrate of silver may be immediately added to a portion of it. If no precipitate appear after the lapse of several minutes, neither chlorides, bromides, iodides, cyanides, sulphides, phosphates, arseniates, arsenites, chromates, silicates, oxalates, nor tartrates can be present. If the original solution contained any considerable quantity of a borate, the borate of silver would be precipitated under these conditions; but a small proportion of some borate might escape precipitation. If the original solution be acid to test paper, add nitrate of silver to a portion of it in a test-tube, and then pour in upon the liquid some dilute ammonia-water, so gently that the two liquids do not mix at once. At some layer near the junction of the two dissimilar liquids, the fluid must be neutral. If at that layer no precipitate or cloud appear, the twelve kinds of salts above enumerated are absent. If the original solution is alkaline, dilute nitric acid is to be added in precisely the same manner as the ammonia-water in the opposite case. The neutral layer between the two liquids is attentively observed, and the absence of any film or cloud therein justifies the same sweeping conclusion as that above given.

If any precipitate is produced by nitrate of silver, its color is to be observed, for some conclusions may often be drawn from this color. Chloride, bromide, cyanide, oxalate, tartrate, silicate, and borate of silver are white; iodide, phosphate, carbonate and arsenite of silver are yellow; arseniate of silver is brownish-red; chromate of silver is purplish-red; sulphide of silver is black. When the silver precipitate is white, black, or of some obscure, indecisive color, the operations in the wet way at this stage should be directed to proving or disproving the presence of chlorides, bromides, iodides, cyanides and sulphides. To this end the portion of the original solution which has been already tested with nitrate of silver, should be made decidedly acid with dilute nitric acid. Of all the silver salts which can be precipitated on the addition of nitrate of silver, only the chloride, bromide, iodide, cyanide and sulphide can resist dilute nitric acid. If the precipitate once formed redissolves completely in nitric acid, no chloride, bromide, iodide, cyanide, or sulphide was present in the original solution. If, on the contrary, a residue remain, one or more of these kinds of salts must have been represented in the original solution. If the residue be black or blackish, the presence of a sulphide is to be inferred ; if it be white or whitish, the absence of sulphides and the presence of a chloride, bromide, or cyanide is to be inferred ; if it be distinctly yellowish, an iodide is probably present. An experiment will best illustrate the most appropriate treatment of a silver precipitate insoluble in nitric acid.

50. *Illustration of the Silver Test.* Prepare in a test-tube a weak solution of chloride of sodium, iodide of potassium, cyanide of potassium, and phosphate of sodium. Add nitrate of silver (§ 131) to this slightly alkaline solution, until the precipitation is complete. The dense precipitate is yellowish-white. Pour dilute nitric acid into the mixture, until the solution is strongly acid ; shake up the contents of the tube thoroughly, and after the lapse of several minutes collect the insoluble precipitate on a filter, and receive the filtered liquid in a test-tube. If the filtrate be neutralized again with ammonia, a yellow precipitate of phosphate of silver will reappear. The precipitate on the filter is then thoroughly washed to free it from the superfluous nitrate of silver. Wash the precipitate into a clean test-tube, decant the water from above it, pour over it ammonia water, and gently heat the mixture. The silver precipitate will visibly diminish in bulk, but a yellowish portion remains undissolved. Filter again, and neutralize the filtrate with nitric acid ; a white

precipitate will fall. This experiment proves that a portion, but not all, of the mixed silver salts which are insoluble in nitric acid, are soluble in ammonia-water. The fact is that the *chloride*, *cyanide* and *bromide* of silver dissolve in ammonia-water, the latter with difficulty, while the *iodide* is insoluble in that reagent.

Special tests, hereafter to be described, are applied in order to confirm the presence of iodine, and to detect each and all of the three substances which are liable to be confounded in the white precipitate just mentioned.

Cyanide of mercury does not give a precipitate with nitrate of silver. When mercury has been detected in the substance under examination, cyanogen must be sought for in other ways (§ 55) than this.

When the liquid under examination contains a ferrous salt, protochloride of tin, or any other active reducing agent, metallic silver is liable to be precipitated as a dark, heavy powder. The examination for the metallic element will generally have revealed the presence of any such agent.

51. *Nitrates, Chlorates, and Acetates.* It is quite clear that no method of precipitation whatever will apply to nitrates, chlorates, and acetates, since these kinds of salts are all soluble in water. No insoluble nitrate, chlorate, or acetate is known. Special tests must be resorted to for these three classes of salts.

CHAPTER X.

SPECIAL TESTS FOR THE NON-METALLIC ELEMENTS.

52. We now proceed to indicate the most available special tests for the non-metallic elements and their commonest compounds. It is noteworthy that the non-metallic elements enter into composition under various forms, which produce with one and the same metallic element various salts. Thus within the narrow range of this treatise, sulphur is to be sought in sulphides, sulphates, sulphites and hyposulphites ; carbon in cyanides, acetates, carbonates, oxalates and tartrates, and arsenic in arsenites and arseniates. The various classes of salts will be taken up successively. It should be premised that these special tests are sometimes applied to the original solution, *before* the precipitation with chloride of barium and nitrate of silver, and sometimes *during or after* these more general testings. In the first case the student is seeking guidance in the application of the more comprehensive tests ; in the latter case he is trying to confirm results already almost sure.

53. *Effervescence.* When a solution containing a carbonate, cyanide, sulphide, sulphite, or hyposulphite, or a mixture of representatives of some or all of these kinds of salts, is treated with chlorhydric acid and then warmed, an evolution of gas occurs with more or less effervescence. The gases evolved are all colorless ; but they all have very characteristic odors except carbonic acid, the gas which escapes from a carbonate. A cyanide gives off the pungent odor of cyanhydric acid. A sulphide yields sulphuretted hydrogen of familiar presence. Sulphurous acid escapes from sulphites and hyposulphites alike ; but in the latter case a deposition of sulphur makes the liquid turbid. If only one of these gases were present, the effervescence and the smell would identify it ; but when mixtures are to be dealt with, further means of identification are necessary.

54. *Carbonates.* To prove the presence of carbonic acid gas, when effervescence has occurred, add chlorhydric acid, little by little, to the effervescing solution, until the acid is decidedly in excess ; meanwhile keep the mouth of the test-tube loosely closed with the thumb to promote the accumulation of the gas evolved. When the tube is supposed to be full, carefully decant the gas into a second test-tube containing a teaspoonful of lime-water (§ 133), taking care not to allow any of the liquid to pass over with the gas. Mix the lime-water and the gas in the second test-tube by thorough shaking. A white precipitate of carbonate of calcium will be produced, if the gas tested is, or contains, carbonic acid.

If the effervescence is slight, and the quantity of gas evolved seems too small to be decanted in this way, dip the end of a dark-colored glass rod into lime-water, and thrust the moistened end into the test-tube, bringing it close to the surface of the fluid. If the gas be carbonic acid, the lime-water adhering to the rod will become visibly turbid.

The student who desires to see the working of this test before having occasion to apply it in an actual analysis, can operate upon a morsel of carbonate of sodium dissolved in a little water.

55. *Cyanides.* When the smell of the gas which escapes from the solution under examination (§ 53), or the qualities of the precipitate with nitrate of silver (§§ 49, 50), give occasion to suspect the presence of a cyanide, the following confirmatory test may be resorted to :—Add to the solution supposed to contain free cyanhydric acid, or an alkaline cyanide, a few drops of a solution containing both a ferrous and a ferric salt (a solution of ferrous sulphate which has been exposed to the air, for example), and a small quantity of caustic soda solution. If cyanogen is present, a bluish green precipitate forms, which consists of a mixture of Prussian-blue and the hydrates of iron. Warm the liquid, and add to it an excess of chlorhydric acid. The hydrates of iron dissolve, but the Prussian-blue remains undissolved.

If only a very small quantity of cyanogen be present, the liquid simply appears green after the addition of chlorhydric acid, and it is only after long standing that a trifling blue precipitate separates from it.

This test may be well illustrated by means of a small particle of cyanide of potassium dissolved in a teaspoonful of water.

To detect cyanogen in cyanide of mercury, it is necessary to precipitate the mercury as sulphide, by means of sulphuretted hydrogen, and to identify the free cyanhydric acid in the filtered or decanted fluid.

56. *Sulphides.* Many sulphides give off sulphuretted hydrogen when heated with chlorhydric acid. If the quantity of gas is so small that its odor is imperceptible, the lead-paper test (§ 31) should be applied.

When sulphides are dissolved in nitric acid or aqua-regia, their sulphur is partly separated in a free state and partly converted into sulphuric acid. The free sulphur is identified by its color and texture, and by its behavior when burnt. The sulphuric acid in the liquid is detected in the usual manner (§§ 45, 46).

57. *Sulphites.* All the sulphites evolve sulphurous acid, without any deposition of sulphur, when treated with chlorhydric acid. The very sharp odor of this gas is enough to identify it.

As has been before stated, nitrate of silver produces in solutions of sulphites a white precipitate ; but this precipitate blackens when the liquid is boiled, on account of the reduction of silver.

When the sulphites are heated with strong nitric acid, or other powerful oxidizing agent, they are converted into sulphates without precipitation of sulphur. The sulphates so produced may be identified in the usual way (§§ 45, 46).

58. *Hyposulphites.* The hyposulphites disengage sulphurous acid and deposit sulphur when warmed with chlorhydric acid. This decomposition is not immediate if the solution be dilute.

The precipitate produced in the solution of a hyposulphite by nitrate of silver, dissolves again readily in an excess of the hyposulphite. On standing, the precipitate of hyposulphite of silver turns black spontaneously, being decomposed into sulphide of silver and sulphuric acid. Heating produces this effect almost immediately.

Hyposulphite of sodium is a good salt from which to get the above reactions of hyposulphites.

59. During the examination for the metallic elements, chromium and arsenic are detected, and the analyst generally obtains pretty certain evidence concerning the

actual condition in which these elements enter into the substance under examination, whether as chromate or salt of chromium, as arsenite or arseniate ; for a chromate is reduced with change of color (§ 22), and sulphide of arsenic is thrown down immediately from an arsenite, but only after long contact with sulphuretted hydrogen from an arseniate. To confirm the indications then obtained, the following tests are used.

60. *Chromates.* When acetate of lead (§ 137) is added to a neutral solution of a chromate, yellow chromate of lead separates, insoluble in acetic acid, but soluble in caustic soda.

As has been before stated, the purplish-red color of chromate of silver (§ 49), and the yellow color of chromate of barium (§ 46), are valuable indications of the presence of a chromate.

A solution of normal chromate of potassium is the best substance from which to obtain for the first time the reactions of chromates.

61. *Arsenites and Arseniates.* Though sure of the presence of arsenic, the analyst is often left in doubt which of these two classes of arsenic compounds he has to deal with. Discriminating tests are necessary.

A *neutral* solution of an arsenite yields with nitrate of silver a yellow precipitate of arsenite of silver. A *neutral* solution of an arseniate gives with the same reagent a brownish-red precipitate of arseniate of silver. Both these precipitates are readily soluble in dilute nitric acid and in ammonia, and they are by no means insoluble in nitrate of ammonium. If solutions containing free arsenious or arsenic acid are to be tested, they must first be cautiously neutralized with the least possible quantity of ammonia-water.

The above tests are generally sufficient, but circumstances may arise in which they would be inapplicable. The following tests afford further means of discrimination :—

If to a solution of arsenious acid or an arsenite, caustic soda be first added in excess, and then five or six drops of a dilute solution of sulphate of copper, a clear bluish liquid is obtained, which, upon boiling, deposits a red precipitate of dinoxide of copper (Cu_2O), while soluble arseniate of sodium is simultaneously produced, and remains in the solution. This test is good as a means of distinguishing between arsenites and arseniates when no organic matters are contained in the solution under examination. The qualification is necessary, because grape-sugar and many other organic substances exercise a like reducing action on copper salts.

If a solution of arsenic acid, or of an arseniate soluble in water, be added to a clear mixture of sulphate of magnesium, chloride of ammonium and ammonia-water (§ 139), a crystalline precipitate of arseniate of magnesium and ammonium will separate, after an interval which is short in proportion to the concentration of the arsenic solution.

62. *Sulphates.* The chloride of barium test (§§ 45, 46) is all-sufficient for the detection of sulphates. The operator should make sure that the chlorhydric acid itself contains no sulphuric acid, that an excess of acid is really present, and that the solution is tolerably dilute. Concentrated acids and strong solutions of many salts impair the delicacy of the reaction.

63. *Phosphates.* When the previous steps of the analysis have proved that the phosphates present in the solution under examination are soluble in ammoniacal liquids, and that no arsenic acid or arseniates are present, the following test will satisfactorily identify a phosphate or free phosphoric acid :—

Add to the solution to be tested a clear mixture of sulphate of magnesium, chloride of ammonium and ammonia-water (§ 139). When a phosphate or free phosphoric acid is present, a white, crystalline precipitate of phosphate of magnesium and ammonium is formed, even in very dilute solutions. Stirring and shaking promote its separation. The precipitate dissolves readily in acids.

When arsenic acid or arseniates are present in the original mixture, this test for phosphates can still be applied, if all the arsenic be previously removed by precipitation as sulphide (§ 23). The magnesium mixture can be used in the filtrate from the sulphide of arsenic.

The preceding test can only be applied when the phosphate present is soluble in ammoniacal solutions. The following test is of much more general application ; it can be used in presence of arsenic acid, and is applicable to either neutral or acid solutions of phosphates ; it is also extremely delicate.

When two or three drops of a neutral or acid solution of a phosphate (even of iron, aluminum, barium, strontium, calcium or magnesium [compare § 27]), are poured into a test-tube containing four or five teaspoonfuls of a solution of molybdate of ammonium in nitric acid (§ 115), there is formed *in the cold* a pale-yellow precipitate which is apt to gather upon the sides and bottom of the tube. If the precipitate does not appear in a few minutes, a few drops more of the solution to be tested may be added. This precipitate is soluble in an excess of phosphoric and other acids ; and certain organic substances also prevent its formation. A yellow coloration of the liquid merely is not enough to prove beyond question the presence of a phosphate ; a precipitate must be waited for. The yellow precipitate can be easily recognized, even in dark-colored liquids, when it has settled. The solution to be tested must not be heated, nor must it be more than blood-warm.

Phosphate of sodium is the best substance on which to try the test for phosphates.

64. *Oxalates.* The precipitation of white, finely divided oxalate of calcium, by all soluble calcium salts, from solutions of oxalates or oxalic acid, has been already described (§ 48). Even the solution of sulphate of calcium gives this reaction with oxalates.

If oxalic acid, or an oxalate in the dry state, be heated in a test-tube with an excess of concentrated sulphuric acid, a mixture of carbonic oxide and carbonic acid is set free with effervescence ; the carbonic acid may be identified by the lime-water test ; and if the quantity operated upon is considerable, the carbonic oxide may be inflamed at the mouth of the tube.

65. *Tartrates.* Tartaric acid and the tartrates, when heated in the dry state char, and emit a very characteristic odor which somewhat resembles that of burnt sugar. This is the only class of salts, among all those within the scope of this treatise, which exhibit this carbonization by heat. Strong sulphuric acid blackens tartaric acid and the tartrates.

To confirm the presence of tartaric acid, or a tartrate, in any liquid supposed to contain it, a concentrated solution of acetate of potassium is added to the liquid, and the mixture violently shaken. The precipitate, when one forms, is a difficultly soluble acid tartrate of potassium.

The addition of an equal volume of alcohol increases the delicacy of the reaction. The more concentrated the solution to be tested, the better. To prepare the required solution of acetate of potassium at the moment of use, rub together in a dish half a teaspoonful of carbonate of potassium and as many drops of acetic acid as will dissolve three-quarters of the carbonate ; throw the mixture on a small moistened filter and use the filtrate.

Tartaric acid is a good substance from which to get these reactions.

66. *Borates.* To confirm the presence of a borate, strong sulphuric acid is mixed with the dry substance under examination in quantity sufficient to make a thin paste, and an equal bulk of alcohol is added to the mixture. The alcohol is then kindled. Boracic acid imparts to the alcohol flame a yellowish-green color. The test is made more delicate by stirring the mixture, and by repeatedly extinguishing and rekindling the flame. Copper salts impart a somewhat similar color to the flame; but this metal, if present, may be got rid of by sulphuretted hydrogen before testing for boracic acid.

If a solution of boracic acid, or of a colorless borate, is mixed with chlorhydric acid to slight but distinct acid reaction, and a slip of turmeric paper (§ 149) is dipped half way into the liquid and then dried at 100° C., the dipped half shows a peculiar red tint. This test is delicate, but there are a few other solutions which impart, not the same, but somewhat similar tints to turmeric paper.

The reactions of borates may be obtained with a fragment of borax.

67. *Silicates.* The silicates of sodium and potassium are the only silicates which are soluble in water. The solutions of these alkaline silicates are decomposed by all acids. If chlorhydric acid is added gradually to a strong solution of an alkaline silicate, the greater part of the silicic acid separates as a gelatinous hydrate. As a rule, the more dilute the fluid, the more silicic acid remains in solution.

If the solution of an alkaline silicate, mixed with chlorhydric or nitric acid in excess, be evaporated to dryness, silicic acid separates ; if the dry mass be ignited and then treated with dilute chlorhydric or nitric acid, the whole of the silicic acid remains insoluble in the free state, as a gritty, whitish powder, while the other substances dissolve.

A solution of chloride of ammonium produces a gelatinous precipitate in strong and moderately dilute solutions of the alkaline silicates. This precipitate is hydrated silicic acid containing alkali.

A solution of waterglass is the best substance in which to study the reactions of the silicates of the alkali-metals.

68. *Fluorides.* If a finely pulverized fluoride is heated in a small leaden capsule or platinum crucible with concentrated sulphuric acid, fluorhydric acid is disengaged.

Coat with wax the convex face of a watch-glass large enough to cover the capsule, by heating the glass cautiously, and spreading a small bit of wax evenly over it while the glass is hot. Trace some lines or letters through the wax with a pointed instrument of wood or horn. Fill the hollow of the glass with cold water, and cover with it the capsule which contains the fluoride mixture. Heat the capsule *gently* for half an hour or an hour. Then remove the watch-glass, dry it, heat it cautiously to melt the wax, and wipe it with a bit of paper. The lines or letters traced through the wax will be found etched into the glass. A barely perceptible etching is made more visible by breathing upon the glass. If much silicic acid is present, this reaction fails.

When a fluoride, naturally combined or artificially mixed with silica, is heated with strong sulphuric acid, fluoride of silicon is evolved. This reaction is available as a test for fluorine.

A mixture of the supposed fluoride and fine dry sand is heated in a short, dry test-tube, with concentrated sulphuric acid. A drop of water, caught in the loop of a clean platinum wire, is held in the mouth of the test-tube. This drop of water becomes merely dim, quite opaque, or almost solid with silicic acid, according to the quantity of fluoride of silicon evolved from the mixture. The gaseous fluoride of silicon shows white fumes when it comes in contact with moist air. If a considerable quantity of fluoride of silicon be evolved from the mixture tested, it can be decanted into another test-tube, and there shaken up with water. If the substance to be tested for fluorine is known to contain silica, it is, of course, unnecessary to add sand to it. This method applies to all fluorides decomposable by hot sulphuric acid. It is evident that this test reversed can be applied to the detection of silica.

Fluoride of calcium (fluor-spar) is a good material from which to obtain these two tests for fluorine.

69. *Chlorides.* The following confirmatory test is applied to chlorides in the dry state :—

When a chloride, in powder, is heated in a test-tube with black oxide of manganese and strong sulphuric acid, chlorine gas is evolved ; this gas is recognized by its odor, greenish-yellow color, and reaction with iodo-starch paper. The gas evolved by a chloride gives no colored reaction with starch alone ; but when a moistened slip of paper, on which a mixture of starch paste and iodide of potassium (§ 130) has been spread, is held in an atmosphere or current of chlorine, the paper is colored blue in consequence of the liberation of iodine which the chlorine effects. The yellow color of the gas is best seen by looking lengthwise through the tube.

70. *Bromides.* The confirmatory tests for bromides depend upon the setting free of bromine itself.

Hot nitric acid liberates the bromine from all bromides except those of silver and mercury. In solutions, the free bromine produces a yellow coloration ; when set free from solid bromides, the brownish-yellow vapors of bromine condense into a liquid upon the cold walls of the tube.

When bromides, in powder, are heated in a test-tube with black oxide of manganese and strong sulphuric acid, brownish-red vapors of bromine are evolved. If chlorides are also present, the bromine will be mixed with chlorine.

To identify bromine and distinguish it from chlorine, moistened starch is brought into contact with the free bromine. A yellow or orange-yellow coloration of the starch marks the presence of bromine. To apply this test, thrust a rod smeared with starch-paste into the tube which contains the bromine vapors ; or, when greater delicacy is requisite, perform the experiment which is expected to liberate bromine in a very small beaker, and cover this beaker with a watch-glass to whose under-side is attached a bit of paper moistened with starch-paste and sprinkled with dry starch.

Bromide of potassium is a good substance with which to study the tests for bromine.

71. *Iodides.* When an iodide, in the solid form or in solution, is heated with strong nitric acid, iodine is liberated and sublimes in violet vapors.

Free iodine in vapor is recognized by the deep blue color which it imparts to starch-paste. Vapors may be tested by bringing into contact with them a glass rod smeared with thin starch-paste, or a slip of white paper on which the paste has been spread.

The best method of detecting iodine in a solution is to add a few drops of thin, clear starch-paste to the liquid, and then set free the iodine by means of nitrite of potassium (§ 129), as follows :—The cold fluid to be tested is acidulated with dilute chlorhydric or sulphuric acid, after the addition of the starch-paste, and a drop or two of a concentrated solution of nitrite of potassium is then added. A dark blue color will be instantly produced. It is essential that the liquid should be kept cool, for the blue coloration is destroyed by heat.

Like chlorine and bromine, iodine is liberated by heating an iodide with black oxide of manganese and sulphuric acid. The iodine so liberated is readily distinguished by the above tests.

The student can try all these tests for iodine with a small crystal of iodide of potassium.

72. *Nitrates.* To confirm the presence of a nitrate, one or both of the two following tests may be used :—

If the solution of a nitrate is mixed with an equal volume of strong sulphuric acid, the mixture cooled in cold water, and a concentrated solution of ferrous sulphate then cautiously added to it in such a way that the two fluids do not mix, the stratum of contact shows a purple or reddish color, which changes to a brown. If the fluids are then mixed, a clear, brownish-purple liquid is obtained. The color fades on heating. Another way of performing the same test is to drop a crystal of copperas into the cold mixture of nitrate and sulphuric acid. There forms around the crystal a dark halo, which disappears with a kind of effervescence on the application of heat. It is, of course, essential that the sulphuric acid employed for this test should be so free from nitric and hyponitric acids, as not itself to give this reaction with ferrous sulphate.

Boil some chlorhydric acid in a test-tube, add to it one or two drops of a dilute solution of sulphindigotic acid (§ 147), and continue the boiling a moment. If the chlorhydric acid is sufficiently free from chlorine, the resulting liquor will be of a faint blue color. If a nitrate, either solid or in solution, be added to this liquid, and the mixture be again boiled, the liquid will be decolorized. This reaction is delicate ; but there are some other substances, especially free chlorine, which have a like bleaching effect.

Nitrate of potassium is a suitable material on which to illustrate the tests for nitrates.

73. *Chlorates.* The preliminary examination gives warning of the presence of chlorates.

When a few particles of a chlorate in the solid form are covered with two or three times as much strong sulphuric acid, and the mixture is *gently* warmed, the liquid becomes intensely yellow, and a greenish-yellow irritating gas of peculiar odor (hypochloric acid, ClO_2) is evolved, which explodes with violence at a moderate heat. After this decomposition, the gas evolved has the characteristic odor of chlorine. The quantity of chlorate operated upon should be very small.

The solution of a chlorate decolorizes indigo-solution precisely like the solution of a nitrate, under like conditions (§ 72).

A chlorate is converted by ignition into chloride, from a solution of which nitrate of silver precipitates the chlorine.

Chlorate of potassium illustrates very well the reactions of chlorates.

74. *Acetates.* When acetates are moderately heated with strong sulphuric acid, hydrated acetic acid distils from the mixture, and may be recognized by its pungent odor.

When an acetate is heated with alcohol and sulphuric acid in equal volumes, acetic ether is formed. The agreeable odor of this ether is highly characteristic.

Hot concentrated sulphuric acid produces no blackening with an acetate.

When a few drops of a solution of ferric chloride (§ 140) are added to a solution of a neutral acetate, or to a solution of an acetate previously neutralized with ammonia, the liquid acquires a dark red color, because of the formation of ferric acetate. If the liquid contain an excess of the acetate, a basic acetate of iron is precipitated in yellow flocks upon boiling, and the fluid finally becomes colorless.

PART SECOND.

PRELIMINARY TREATMENT.

THE ORDER OF PROCEDURE.

75. The substance to be examined may be either solid or liquid. We shall consider first the preliminary treatment of a solid ; afterwards that of a liquid. The solid may be a metallic substance, that is, a pure metal or an alloy, or it may be a salt, mineral, or other non-metallic body. The method of procedure differs in the two cases. We shall describe first the treatment of a salt, mineral, or other non-metallic substance. The two following observations, however, apply to all cases. The student should, in the first place, learn as much as possible from the external properties of the substance to be analyzed, from its color, consistency, and odor, if it is a liquid ; from its color, texture, odor, lustre, hardness, gravity, and crystalline or amorphous structure, if it is a solid. By attentively observing the characteristics or individual peculiarities of every substance which passes through his hands, the student will soon learn to recognize many substances at sight,—by far the quickest and easiest way of identifying them. Secondly, since the original substance must be several times reverted to in order to complete an analysis, the student should husband his stock of the substance to be analyzed, never employing the whole of it for any single course of experiment. It is well also to reserve a portion for unforeseen contingencies.

CHAPTER XI.

TREATMENT OF A SALT, MINERAL, OR OTHER NON-METALLIC SOLID.

A. PRELIMINARY EXAMINATION IN THE DRY WAY.

76. *Closed-tube Test.* Prepare a hard glass tube No. 4 (§ 171), about 3 inches long, and closed at one end. Let fall into this tube a minute fragment of the solid, or a little of its powder. If the substance be used in powder, wipe out the tube with a tuft of cotton on a wire, in order that the interior walls of the tube may be clean to receive a sublimate. Heat the substance at the end of the tube, at first gently in the lamp, but finally intensely in the blow-pipe flame. The following are the most noteworthy reactions with the inferences to be drawn from them ; it not unfrequently happens that a single substance gives several of these reactions :—

I. *The substance blackens,* and gases or vapors are evolved. These vapors often have a disagreeable smell, sometimes like that of burnt sugar, paper, or feathers. Sometimes they condense in tarry droplets ; water also condenses on the cold part of the tube. These appearances indicate the presence of organic substances.

Now, the presence of fixed organic matter interferes

with the detection of many substances, and it must be destroyed before the analysis can be proceeded with. A portion of the original substance, sufficient for the regular course of examination for the metallic elements, is ignited in a porcelain crucible, with free access of air, or on platinum foil, if the presence of no metal be suspected, until all the carbon is burnt out of it. This ignition is best performed on successive small portions rather than on a large mass at once.

It is obvious that some inorganic volatile matters may be lost during this ignition. Furthermore, some substances, especially alumina and chromic and ferric oxides, are made very insoluble by ignition. Exceptionally, therefore, the following process, which is not liable to these objections, is employed:—The substance in powder, paste, or concentrated solution, is heated in an evaporating-dish, with strong nitric acid, to a temperature just below boiling. To this hot mixture chlorate of potassium, in small bits, is added gradually, until the organic matter is all destroyed. The solution is then evaporated to dryness on a water-bath; the dry residue is moistened with strong chlorhydric acid, the mixture diluted with water, warmed and filtered, if there be any residue. The filtrate is fit for the regular course of analysis, except that potassium, having been added, must not be tested for in this liquid. The residue, if any, must be examined for the insoluble chlorides of Class I. This process is simply a combustion at a low temperature.

Simple blackening is not proof of the presence of organic bodies. Some salts of copper and cobalt, for example, blacken through the formation of a black oxide.

When organic matter is shown to be present, the student should look particularly for acetic (§ 74) and tartaric (§ 65) acids; but he will not forget that there are hundreds of organic acids which are not comprehended in the plan of this treatise.

II. The substance does not carbonize, but vapors or gases escape from it. The most important are :—

a. Aqueous vapor, which condenses in the upper part of the tube. Test this water with litmus paper ; if it is alkaline, ammonia may be suspected ; if acid, some volatile acid (H_2SO_4, HCl, HBr, HI, HFl, HNO_3, &c).

b. Oxygen, recognized by its relighting a glowing match. This gas indicates nitrates, chlorates, and peroxides. If the heated substance fuses, and a small fragment of charcoal thrown in is energetically consumed, the presence of a nitrate or chlorate may be assumed.

c. Hyponitric acid, recognized by the brownish-red color of the fumes. It results from the decomposition of nitrates.

d. Sulphurous acid, recognized by its odor. It not unfrequently results from the decomposition of sulphates, sulphites, and sulphides.

e. Carbonic acid, derived from decomposable carbonates, and to be recognized by lime-water (§ 54).

f. Cyanogen, derived from decomposable cyanides, and to be recognized by its odor, and the blue flame with which it burns when there is enough of it to be lighted.

g. Sulphydric acid gas, derived from moist sulphides and to be known by its smell.

h. Ammonia, resulting sometimes from the decomposition of ammoniacal salts.

III. A sublimate forms beyond the heated portion of the tube. The whole of the substance may volatilize. The following are the commonest sublimates :—

a. Sulphur, which sublimes in reddish drops. The sublimate becomes solid and yellow, or yellowish-brown, on cooling.

b. Ammonium-salts give white sublimates. Test a separate small portion of the original substance for the salts of ammonium, by mixing it in a small test-tube with an equal bulk of slaked lime and a few drops of water, and heating the mixture. Ammonia, when evolved, may be recognized by its smell, and by the white fume produced when a rod, moistened with a mixture of equal parts of strong chlorhydric acid and water, is held above the mouth of the tube. Unless the original solid is obviously inalterable by heat, it should be *invariably* tested in this way for ammonium-salts.

c. Metallic mercury and some of its compounds. The metal sublimes in metallic droplets. The two chlorides of mercury give sublimates which are white when cold. The red iodide of mercury gives a yellow sublimate. The sulphide of mercury gives a dull black sublimate.

d. Arsenic and some of its compoudns. Metallic arsenic gives a black sublimate of metallic lustre. Arsenious acid gives a white sublimate, which looks crystalline under a magnifying lens. The sulphides of arsenic give sublimates which are brownish-red while hot, but reddish-yellow to red when cold. These sublimates look not unlike that of pure sulphur.

e. Teroxide of antimony first fuses to a yellow liquid, and then gives a white sublimate, composed of needle-like crystals.

f. Oxalic acid gives a white, crystalline sublimate, with dense fumes in the tube.

The inferences to be drawn from this simple preliminary experiment are of very unequal value. Thus the detection of organic matter is of the first importance, because, as has been seen, such matters must be got rid of before the analysis can be proceeded with. Again, nitrates and chlorates should be detected with a good degree of certainty by their reactions in the closed tube. Thirdly, the presence, or entire absence, of ammonium-salts should be put beyond doubt at this first stage of the examination. Fourthly, the presence of mercury, or of mercurous salts, determines the choice of the acid solvent in favor of nitric acid, in case water will not dissolve the substance under examination (§ 81), and the presence of mercuric salts renders necessary the substitution of sulphydrate of ammonium for sulphydrate of sodium as the solvent for the sulphides of Class III (§ 43). Accordingly, it is useful to get information of the presence of mercury or its compounds at this early stage of the examination. As to the other appearances, they give information which may be convenient, but is never essential for the safe conduct of the regular course of analysis. They have been described because they may occur with or instead of the really important reactions.

77. *Reduction Test.* Mix a little of the powder of the substance under examination (the bulk of a hemp-seed) with an equal quantity of carbonate of sodium, and make the mixture into a pasty ball with a small drop of water. Select a piece of dry, well-burned, soft-wood charcoal, and cut out of it a rectangular block about 6 inches long, 1¼ in. wide, and ½ to ¾ in. thick, having its

flat, smooth surface (6 in. by $1\frac{1}{4}$ in.) at right angles to the rings of growth in the tree. It is this surface which is always to be used. A good piece of charcoal may be made to serve for many assays, by filing off the used surface and exposing a new one. At a quarter to half an inch from the end of such a piece of charcoal, scoop out with a penknife a little cavity of the size of half a pea. Place the prepared pellet in this cavity, and expose it for several consecutive minutes to the reducing flame of the blow-pipe (§ 167).

Under these conditions, vapors of characteristic odor or appearance may be evolved; some of them will be mentioned below. The two objects, however, to which attention is specially to be directed are the residue in the cavity and the incrustation on the charcoal outside of the cavity.

The following metals may be found as fused metallic globules in the cavity; lead, silver, and gold are reduced with ease, even by an inexperienced operator; tin and copper with some difficulty :—

a. Gold—a yellow, malleable globule, produced without incrustation.

b. Copper—a red, malleable globule, produced without incrustation.

c. Tin—a bright, white, malleable globule. An incrustation is simultaneously produced, which is faint yellow when hot, and white when cold; it immediately surrounds the globule.

d. Lead—a very fusible and very malleable globule. A yellow incrustation is simultaneously produced.

e. Silver—a brilliant, white, malleable globule, produced without incrustation.

Two other common metals, bismuth and antimony, may be reduced to gray metallic globules; but these globules are brittle, and are not liable to be confounded with the malleable globules just described. Bismuth gives a yellow incrustation which resembles that of lead.

Common charcoal is itself very apt to show a grayish incrustation of ash round about the heated assay; this incrustation remains unaltered or increases, when directly exposed to the flame. The student should test each piece of charcoal before the blow-pipe flame, in order that he may not imagine a deposit of ash to be an incrustation derived from the substance under examination.

If a distinct globule has been obtained, it must be picked out with a pair of jewellers' tweezers, and pounded on some smooth and hard body to test its malleability. If it is malleable, replace it upon the charcoal at an unused spot, and heat it strongly with the oxidizing flame. Gold and silver globules fuse, but maintain their brilliancy and give no incrustation; this proof distinguishes a genuine gold globule from a yellow globule composed of an alloy of copper and some white metal. A yellow globule composed of oxidizable metals tarnishes instantly in the oxidizing flame. A tin globule fuses, but its bright surface is instantly tarnished, and a white incrustation of binoxide of tin is produced which cannot be driven off by either flame. A lead globule is rapidly converted into litharge, a yellow incrustation being produced, which volatilizes with a bluish color when touched with the reducing flame. A copper globule is blackened by the formation of oxide of copper, and the blow-pipe flame is tinged with green.

Certain other phenomena may manifest themselves during the experiment for the reduction of malleable metallic globules. Sulphur, ammonium-salts in general, the chlorides, bromides, iodides, and sulphides of sodium and potassium, the chlorides of lead, bismuth, tin and copper, metallic mercury, arsenic, antimony and zinc, and many compounds of these four elements, are liable to pass off in vapors, which are often in part deposited upon the coal at a greater or less distance from the hot assay according to their volatility. With the exception of sulphur, these sublimates are white, but when deposited in a thin film upon the black coal they have a gray or blue appearance. During the production of the arsenic sublimate a peculiar odor is evolved; this sublimate being very volatile is only deposited at a considerable distance from the assay. The incrustation produced by zinc is distinctly yellow while hot, but turns white on cooling; it settles near the assay and is driven away again with difficulty.

These phenomena are all of secondary importance; the main object of the experiment is the reduction of the five malleable metals above enumerated. We may thus obtain knowledge of the presence of gold (for a confirmatory test, see § 92, *b*), a metal not included in our scheme of analysis in the wet way. Tin generally gives warning of its presence during this experiment; and this warning is of use, because it is inconvenient to apply nitric acid as a solvent to a substance containing tin, since this reagent converts tin into the very insoluble binoxide of tin. The detection of copper at this stage is of little advantage. The most important fact deducible from the reduction-test is the presence of either silver or lead.

In dissolving an unknown substance which proves to be insoluble in water, it is customary to try, as the second solvent, chlorhydric acid. The chlorides of silver and lead being insoluble, or difficultly soluble, this acid should not be used as a solvent when either of these two metals is present. Nitric acid must be used instead.

B. DISSOLVING A SALT, MINERAL, OR OTHER NON-METALLIC SOLID, FREE FROM ORGANIC MATTER.

78. Before a solid substance can be submitted to the systematic course of analysis, it must be brought into solution. There is no universal solvent. Different substances require different solvents. The four solvents employed in qualitative analysis for salts, minerals, and other non-metallic solids, are water, chlorhydric acid, nitric acid, and aqua-regia ; and these four liquids are invariably to be tried in the precise order in which they here stand. Water is always to be tried first ; to whatever resists water, strong chlorhydric acid is applied ; if chlorhydric acid fails to dissolve the solid completely, or is unsuitable, for reasons disclosed in the preliminary examination as just explained, nitric acid is tried, and after nitric acid, aqua-regia as the last resort. A solid substance should invariably be reduced to a very fine powder, before being submitted to the action of solvents (§ 180).

79. *Dissolving in water.* About half a thimbleful of the powdered substance is boiled with ten times as much water in a test-tube. If an effervescence occurs, as is possible with mixtures containing an acid salt (yeast-powders, for example), the gas evolved should be carefully tested (§§ 53–58). If the substance dissolves completely, the solution is ready for analysis. When undissolved powder remains in the tube after protracted boiling, filter a few drops of the liquid, and evaporate a drop or two of the filtrate to dryness on clean platinum foil, at as low a heat as possible. If there be no residue on the foil, or if the residue be scarcely appreciable, the substance is practically insoluble in water, and acids must be tried as solvents. But if, on the contrary, a tolerable residue remains on the foil, decant the liquid in the tube into the filter, and boil the powder again with water. Persevere with this treatment until it is evident that a part of the powder is insoluble in water. The insoluble residue in the test-tube is thrown upon the filter; the clear liquids which have passed through the filter, collected together and concentrated by evaporation if of unreasonable bulk, are ready for the regular course of analysis. In this case, and in the still more favorable case in which all the substance has dissolved in water, it is a simple aqueous solution which is submitted to analysis.

80. *An aqueous solution.* If the student is assured that the unknown substance is a simple salt, he may draw some trustworthy inferences from the fact that the substance dissolves in water. Of the salts which fall within the scope of this manual, the following are practically soluble in water :—

1. *All* salts of sodium, and all of potassium and ammonium, except their double platinum-chlorides.
2. *All* nitrates, chlorates, and acetates.
3. Chlorides, bromides, and iodides, except those of silver and mercury. (Mercuric chloride is soluble. The lead salts are difficultly soluble.)
4. Sulphates, except those of barium, strontium, and lead.

5. Many hyposulphites.

6. The sulphides of sodium, potassium, ammonium, magnesium, barium, strontium, and calcium.

7. A few cyanides, oxalates, tartrates, and chromates, besides those of the alkali-metals already mentioned.

It is obvious that any element of the thirty-six considered in this treatise, may be present in an aqueous solution ; but it is also evident from the above list, that a great number of salts are absolutely excluded because of their insolubility in water.

If, on the contrary, there is no certainty that the substance under examination is not a complex artificial mixture, no safe conclusions can be drawn from the fact that a part of it, or the whole of it, dissolves in water.

Test the solution with litmus paper. The solution is either neutral, acid, or alkaline.

Neutral. The normal salts of most of the metals have an acid reaction. Sodium, potassium, barium, strontium, calcium, magnesium, manganese, and silver are the only metallic elements which form salts whose solutions are neutral. Add to two or three drops of the solution a drop or two of carbonate of sodium. If a precipitate forms, any of the above-mentioned elements may be present ; but if no precipitation ensues, only sodium and potassium can be present.

Acid. The acidity may be caused by a normal salt having an acid reaction, or by an acid salt. Neither carbonates nor sulphides can be present in an aqueous solution with an acid reaction.

Alkaline. The alkalinity may be due to the hydrates, sulphides, cyanides, or carbonates, of the metals belonging to Classes VI and VII; to the presence of a borate, silicate, phosphate, arseniate, or aluminate of sodium or potassium; to free ammonia, or carbonate of ammonium. If the alkalinity proceed from an alkaline sulphide, the metals whose sulphides are insoluble in water, and alkaline sulphides, must be absent. If it is due to the presence of the hydrates or carbonates of the metals of Classes VI and VII, a very large number of substances are excluded. If it proceed from ammonia or carbonate of ammonium, all substances precipitable by these reagents are absent.

An alkaline solution may obviously contain some substance, soluble in an alkaline solvent like caustic soda or sulphydrate of ammonium, but liable to immediate precipitation when this solvent is destroyed by the addition of chlorhydric acid at the first step of the analysis. Thus any sulphide of Class III dissolved in caustic soda, or an alkaline sulphide, or compounds of alkaline hydrates with the hydrates of aluminum, zinc or chromium, would be precipitated when the alkaline solvent was neutralized. Chloride of silver dissolved in ammonia-water would be thrown down by any acid added in excess. Again, the alkaline solution of a silicate of sodium or potassium, when neutralized with acid, yields a very gelatinous, whitish precipitate of hydrated silicic acid. From a very concentrated solution of a borate, boracic acid separates in colorless, shining, flat crystals, when the solution is acidified with chlorhydric acid; but the boracic acid thus separated dissolves when the solution is diluted.

In view of these possibilities, an alkaline aqueous solution should be carefully neutralized with nitric acid, as a preliminary measure, before chlorhydric acid is added to it. Effervescence should be watched for, and if it occurs, studied as directed in §§ 53–58. Several different cases of precipitation may be distinguished, requiring somewhat different treatment.

a. If the characteristic gelatinous precipitate of silicic acid appears, the acidulated solution must be evaporated to dryness and ignited. The silicic acid is thus rendered insoluble. The ignited residue is digested with dilute nitric acid and filtered. The filtrate is ready for the usual course of analysis. The insoluble residue is silicic acid.

b. If the glistening, colorless plates of boracic acid appear, dilution with warm water will cause them to redissolve.

c. If a precipitate appear on neutralization, whose color or texture proves that it is neither silicic nor boracic acid, but some substance insoluble in water and dilute acids, thrown down in consequence of the destruction of its alkaline solvent, the liquid is made slightly acid, and then filtered. The filtrate is ready for the usual course of analysis. The precipitate, rinsed with a little water, is reserved for further treatment; it is not properly a substance soluble in water, and it must be brought into solution by other methods, hereafter to be described. Sometimes a precipitate forms on exact neutralization of an alkaline fluid, which redissolves when the acid is added in excess.

81. *Dissolving in acids.* The substance which water has failed to dissolve, either in whole or in part, is next boiled in a small dish with three or four times its bulk of concentrated chlorhydric acid, *unless* the tube-test (§ 76) or the reduction-test (§ 77) has proved the presence of silver, lead, or mercury; in which case nitric acid is the first acid to be tried. (See the next paragraph.) If an effervescence occur, the escaping gas is to be tested, as described in §§ 53–58. After boiling the powdered substance with the strong acid, dilute the fluid with twice its bulk of water, and repeat the boiling if any residue remain undissolved. The acid is diluted because, though the substance to be dissolved is best attacked in the first instance by strong acid, the salts formed by the action of the concentrated acid are more likely to dissolve readily in a dilute than in a strong acid. Not a few salts which scarcely dissolve in strong acids, are readily soluble in the same acids when diluted. If the whole of the substance finally dissolves, the solution still farther diluted is ready for the transmission of sulphuretted hydrogen (§ 22), for it is of course unnecessary to examine it for members of Class I. If, on the contrary, an undissolved residue remain in the tube, ascertain if anything has dissolved, by carefully evaporating two or three drops of the fluid to dryness on platinum foil. Should an appreciable residue, in excess of that given by two or three drops of the acid employed, remain upon the foil, separate the liquid in the tube from the undissolved substance by decantation or filtration. Reserve the solution, labelling it "HCl Sol."

Rinse the undissolved powder with water, and then boil it in an evaporating dish with three or four times its bulk of strong nitric acid. If the original substance contained silver, lead, or a mercurous salt, chlorhydric acid will not have been used, and it will be the residue from the aqueous solution, which is now to be boiled with nitric acid. In this case, effervescence is to be watched for. If the substance dissolves completely in the strong acid, or dissolves, with the exception of a light yellow mass of sulphur, which often separates from a sulphide, evaporate the liquid to a very small bulk to drive off the free acid, dilute the evaporated solution with several times its bulk of water, separate the sulphur, if necessary, by filtration, and reserve the solution, labelling it "HNO_3 Sol." If the substance does not completely dissolve in the strong acid, dilute the fluid with twice its bulk of water, and repeat the boiling. If the dilute nitric acid effects complete solution, reserve the solution, labelling it as before, "HNO_3 Sol." If neither the strong nor the diluted nitric acid effects the complete solution of the substance, ascertain if anything has dissolved in the dilute acid by the usual test on platinum foil. If an appreciable residue remain on the foil, separate the undissolved solid in the dish from the liquid by decantation or filtration, and reserve the solution.

Boil the powder, which has resisted both acids taken singly, with aqua-regia. If it dissolves completely, evaporate the solution to a very small bulk, dilute the evaporated solution largely with water, and reserve it for analysis, labelling it "Aq. reg. Sol." It is useless to look for members of Class I in such a solution. If, on the contrary, it does not completely dissolve after protracted boiling, test the liquor to see if anything has dissolved. If an appreciable residue remains on the foil, dilute the acid fluid, filter it, reserve the solution labelled as before, and wash the undissolved residue thoroughly with water, to prepare it for further treatment (§ 83).

82. *An acid solution.* Of the three kinds of acid solutions here described, any one, any two, or all three, may be obtained from a single mixture of different solids. There is an advantage in knowing that a part of a complex mixture is soluble in water, a part in chlorhydric acid, a part in nitric acid, and a part only in aqua-regia ;

because this knowledge may enable the student, when he has found out all the elements of the mixture, to make a more probable guess at the manner of their combination in the original mixture than he would otherwise be able to. But it is quite unnecessary to keep the three kinds of acid solution apart, when all three have been obtained, and to analyze them separately. On the contrary, all three should be mixed together, and analyzed in one course of testing. It must only be borne in mind that when lead, silver, or mercurous salts are present, the nitric acid solution of the residue from the aqueous solution will give a precipitate of the insoluble chlorides of Class I, on being mixed with a chlorhydric acid or aqua-regia solution.

The student must be careful to use no more acid than is absolutely essential. Nitric acid, particularly, is very objectionable; because, when free, it reacts upon sulphuretted hydrogen with mutual decomposition, sulphur being set free.

Sometimes a strongly acid solution becomes turbid when merely diluted with water. This phenomenon points to the presence of bismuth or antimony. The turbidity will disappear again on the addition of chlorhydric acid.

Certain silicates, when boiled with concentrated acids, are decomposed, and gelatinous silicic acid is separated. This happens but rarely, however, in the rapid processes of qualitative analysis; and if it should happen, it is not likely to lead the student into error. A residue insoluble in all acids will remain; this residue is, or contains, free silicic acid.

It must not be supposed that it is common to try all four solvents on one and the same substance. Water and chlorhydric acid are the common solvents; nitric acid and aqua-regia are, actually, but seldom resorted to as solvents, except for metals (§ 92). It would require some ingenuity to devise an artificial mixture which would put to the test all the capabilities of the above-described method of bringing solids into solution in water and acids. Such mixtures are not met with in ordinary experience. It is the object of any method of analysis to meet real problems, not artificial complications which may be imagined, but which do not occur in fact.

C. TREATMENT OF INSOLUBLE SUBSTANCES.

83. The substances of common occurrence which are practically insoluble in water and acids are :—

The sulphates of barium, strontium, and lead.

Chloride of silver.

The anhydrous sesquioxides of aluminum, chromium, and iron, either native or the result of an intense ignition.

Chrome-iron-ore, a native mineral.

Some aluminates.

Binoxide of tin, native, or the result of ignition.

Silica and many silicates.

Fluoride of calcium (fluor-spar).

Besides the substances included in this list, sulphur and carbon, or graphite, should, perhaps, be mentioned, because they are insoluble ; but they will have been detected during the preliminary blow-pipe examination, and their presence allowed for. Bromide, iodide and cyanide of silver, are decomposed by boiling with aqua-regia, and converted into the chloride, so that these substances never appear in their proper form in the final insoluble residue.

84. Substances which resist solution in liquids are generally liquified by the action of fluxes at a high temperature; they are fused in contact with some powerful decomposing agent, like the carbonate or acid sulphate of an alkali-metal, or the hydrate or carbonate of an alkaline-earth metal. Certain preliminary experiments should precede the fusion.

The insoluble powder is first examined carefully (with the help of a lens, if convenient), to ascertain if it is a homogeneous substance of the same color throughout, or a mixture composed of dissimilar, variously-colored particles. The following blow-pipe experiments sometimes give decisive indications, particularly with homogeneous substances:—

a. The reduction-test (§ 77) is repeated with great care, looking especially for silver, lead, and tin, and applying to the globule, if any is obtained, the test for distinguishing between these three white metals. This test has already been applied to the original substance; but if this substance was a complex mixture containing soluble ingredients, it is quite possible that the test should give a more satisfactory result, now that all substances soluble in water and acids have been removed, than it yielded before. If *any* reducible metal is detected, it is necessary to use a porcelain crucible for the fusion which it may be desirable to make (§ 85), in order to convert the insoluble substance into a more manageable form. A platinum crucible, which is employed for most fusions, cannot be used with safety when the substance to be fused contains any reducible metal; for many of the alloys of platinum are readily fusible.

Sometimes, when the substance under examination contains but a small proportion of metal, some metal may be reduced during the blow-pipe experiment on charcoal, but the detached particles may not run together into a single conspicuous globule. Since a mistake as to the presence of a reducible metal may involve the destruction of a platinum crucible, it is best, in doubtful cases, to operate in a more delicate fashion. To ascertain, beyond question, whether any reduced metal has been separated in this experiment, moisten the cavity in the charcoal with water after the fusion has been finished, cut the charcoal out for a little distance, both around and below the cavity, and transfer the contents of the cavity and the scraps of charcoal to an agate or porcelain mortar. Pulverize the whole mass, and then carefully wash away the powdered charcoal, and all the lighter portion of the mixture. Any malleable metal that may have been reduced remains in the mortar in little flattened grains or spangles, in which the peculiar color and lustre of the metal or alloy are generally visible. Sometimes metallic streaks are produced on the mortar or pestle by little particles of metal ground between them. The student must not mistake glistening particles of wet charcoal sticking to the mortar or pestle for metallic spangles.

b. Prepare another pellet of a mixture of equal parts of the insoluble powder and carbonate of sodium, adding a little charcoal powder to the paste. Fuse this mixture upon charcoal in the reducing flame of the blow-pipe. Scoop out the fused mass and the surrounding charcoal with a penknife, place the dry mass upon a bright surface of silver (coin or foil), and wet it with a drop of water. If a brown stain be produced on the silver, it is evidence of the presence of sulphide of sodium in the fused mass. This sulphide results from the reduction of a sulphate, and is evidence of the presence of a sulphate in the substance tested. The odor of sulphuretted hydrogen is often perceptible when the fused mass is moistened. The silver coin or foil may be replaced by a piece of lead-paper, if care be taken not to mistake the mere dirtying of the paper for a stain of sulphide. It is obvious that the carbonate of sodium used in this test must be so free from sulphate of sodium as not itself to give this reaction on silver, after fusion on charcoal. Since coal-gas invariably contains traces of sulphur compounds, this test cannot be performed with a gas-flame.

c. Make the loop on the end of the bit of platinum wire (§ 168) white-hot in the blow-pipe flame, and thrust it white-hot into some powdered borax ; a quantity of borax will adhere to the hot wire ; reheat the loop in the oxidizing flame ; the borax will puff up at first, and then fuse to a transparent glass. If enough borax to form a solid, transparent bead within the loop does not adhere to the hot wire the first time, the hot loop may be dipped a second time into the powdered borax.

When a transparent glass has been formed within the loop of the platinum wire, touch the bead of glass while it is hot and soft to a few particles of the insoluble powder, and reheat the bead with the adhering powder in the oxidizing flame. If the substance dissolves slowly in the borax, and the bead has a fine yellowish-green color when cold, chromium is probably present. Reheat the bead in the reducing flame. If it presents a bright green color both when hot and cold, there is no doubt of the presence of chromium.

It sometimes happens, when too much of the substance to be tested has been added, that the borax bead becomes so dark-colored as to be practically opaque. It may then be flattened, while soft, by sudden pressure between any smooth metallic surfaces, like the flat parts of jewellers' tweezers. If the flattening makes the color of the borax-glass visible, nothing more is necessary; but if the glass is still too dark, all the glass outside the loop of platinum may be broken off by gentle hammering, and the remaining glass may be reheated and largely diluted by the addition of more borax.

It is convenient to be informed of the presence of chromium, because chromic oxide and chrome-iron-ore are substances which it is particularly difficult to decompose effectually by fusion. In presence of chromium, no other bead reaction which can be anticipated under the circumstances will give a decisive result; but in the absence of chromium, the presence of iron may be determined. A suitable quantity of oxide of iron causes the borax-bead, heated in the oxidizing flame, to look red when hot and yellow when cold. In the reducing flame the iron bead becomes greenish, or light brownish-green.

d. The test for fluorine (§ 68) should be applied to the original substance, if it has not already been done.

When all the above-mentioned tests (*a–d*) give negative results, the simplification of the problem is very conspicuous; the substances which may be present are reduced to alumina and some aluminates, silica and silicates. Again, when some of the preliminary tests give affirmative results, the evidence may be almost conclusive, if the substance under examination be evidently homogeneous. Thus chloride of silver, sulphate of lead, chromic or ferric oxide, binoxide of tin, or fluoride of calcium, may be satisfactorily identified.

There are two methods of changing insoluble substances into more manageable forms by the application of heat, with sufficient exactness for the purposes of the qualitative analyst,—the method by fusion, and the method by deflagration.

85. *Fusions.* Mix the fine powder of the insoluble substance with about four parts by weight of dry carbonate of sodium in powder. Both powders must be as fine as they can be made, and they must be intimately mixed. Keep the mixture at a bright red heat, in a platinum crucible (a *porcelain* crucible if a reducible metal has been found in the substance, and fusion is for any reason preferred to deflagration), until the mass has been brought to a state of quiet fusion (§ 164). Place the hot platinum crucible, when withdrawn from the lamp or fire, on a cold block, or thick plate of iron, and let it cool. When the crucible has been cooled in this way, the fused mass can generally be removed from the crucible in an unbroken lump. Soak the lump in boiling water until everything is dissolved which is soluble in water. If the mass cannot be detached from the crucible, the crucible and its contents must be soaked in boiling water.

When the green borax bead, and the dark color of the insoluble powder, point to the presence of chrome-iron-ore, a mixture of two parts, by weight, of carbonate of sodium with two parts, by weight, of nitre, may be substituted for the four parts of carbonate of sodium alone.

86. *Treatment of the fused mass.* The aqueous solution of the fused mass is filtered from the residue insoluble in water. Small portions of it are to be tested separately at this stage of the process, for sulphate, chromate, chloride and fluoride of sodium, either of which salts (besides others not regarded for the moment) may result from the decomposition of the insoluble substance,

and be found in the aqueous solution. A chromate colors the solution yellow.

a. Acidify a small portion with chlorhydric acid, and test for sulphates with chloride of barium (§ 62). The student must learn by trial how much sulphate, if any, his carbonate of sodium contains.

b. Acidify another small portion with acetic acid, and apply the acetate of lead test for chromates (§ 60). In presence of sulphuric acid this test will be obscured, but not rendered wholly useless (§ 29, p. 49).

c. Acidify a third portion with nitric acid, and apply the nitrate of silver test for chlorine (§ 69). The student must first prove that his carbonate of sodium contains no chlorine.

d. A fourth portion, having been concentrated by evaporation in a porcelain dish, and again cooled, is acidified with chlorhydric acid, and then left at rest until the carbonic acid has escaped. It is then supersaturated with ammonia, heated, and filtered while hot. The filtrate is collected in a bottle ; chloride of calcium is immediately added to it ; the bottle is closed and allowed to stand at rest. If the original substance contained a fluoride, the fluorine will have combined with sodium during the fusion, and fluoride of sodium will be contained in the aqueous solution. The carbonic acid having been expelled, and all substances precipitable by ammonia having been removed, the chloride of calcium will throw down the fluoride of calcium. If a precipitate separates from the liquid in the bottle after some time, it is collected in a small filter, dried and examined for fluorine by the method of § 68.

The rest of the aqueous solution is acidified with chlorhydric acid, evaporated to dryness and ignited; the residue thus obtained is boiled with dilute chlorhydric acid. If the dilute acid fails to dissolve the residue completely, the insoluble portion consists of silicic acid. The solution is reserved.

That portion of the original fused mass which boiling water did not dissolve, and which was filtered from the aqueous solution, is next dissolved in acid—chlorhydric acid, if silver and lead be absent; nitric acid, if either of these metals be present—and the solution so obtained, mixed with the reserved solution of the last paragraph, is examined in the usual way for the metallic elements (§ 43), except, of course, sodium (and sometimes potassium), which has been added in the flux.

If a portion of the fused mass resist both water and acids, the insoluble portion may consist of separated silicic acid, or of some of the original substance undecomposed by the fusion. In the latter case, another and more prolonged fusion is the only effectual remedy, although it may often happen that a partial decomposition of the insoluble substance will enable the analyst to recognize all the elements which it contains.

87. *Fusion with Acid Sulphate of Sodium.* The following method may be tried to advantage upon ferric oxide, chromic oxide, or chrome-iron-ore, and some very refractory silicates. Heat the insoluble substance with three or four times its bulk of acid sulphate of sodium (§ 122), in a platinum crucible, until the sulphate melts; then maintain it in the liquid state for half an hour. This operation should be performed under a hood. The fused mass is treated essentially as before (§ 86), allowance being made for the different nature of the flux.

88. *Silicates* are by far the commonest of insoluble substances. A great variety of metallic elements occur in insoluble silicious minerals, so that the possible contents of the acid solution of § 86 are very various. When silicic acid has been found in the aqueous solution of § 86, the acid solution also should be evaporated to dryness, after it has been once made, and the residue should be ignited ; the residual mass is again treated with dilute chlorhydric acid. Silicic acid is again left behind in this process ; and the subsequent examination goes on the better for this preliminary removal of silica, which, if left in solution, might create confusion by appearing as a precipitate at almost any stage of the analysis.

Many silicates contain sodium and potassium. When the presence or absence of these alkali-metals is to be determined, it is evident that the pulverized silicate must not be fused with carbonate of sodium, but with some decomposing flux free from alkali.

89. *Fusion with Carbonate of Calcium and Chloride of Ammonium.* An intimate mixture is prepared of one part of the silicate, six parts of pure precipitated carbonate of calcium, and three-fourths part of pulverized chloride of ammonium. This mixture is heated to bright redness in a platinum crucible for thirty or forty minutes. The crucible, with its contents (which should be in a coherent, sintered, but not thoroughly fused condition), is then placed in a beaker, and soaked for half an hour in water kept near the boiling point. The contents of the beaker are then filtered. The filtrate, containing caustic lime, chloride of calcium, and all the sodium and potassium of the original silicate as chlorides, is treated with a little ammonia-water, and with carbonate of ammonium in slight excess ; the liquid is heated to boiling and filtered. This second filtrate is evaporated to dryness, and gently ignited to expel the ammonium salts. The residue is dissolved in a little water ; one or two drops of carbonate of ammonium, and a drop of oxalate of ammonium, are then added ; the mixture is again heated and filtered ; this third filtrate is evaporated to dryness and ignited ; the ignited residue, if there be any, consists of the chlorides of sodium and potassium, or of one of these two salts. This residue is examined according to § 41.

90. *Deflagration.* The method of fusion just described involves the use of a platinum or porcelain crucible, and demands the heat of a blast-lamp, or strong coal fire. Neither crucibles, lamps, nor fires are necessary in the method of deflagration, which applies the heat inside the mass to be fused. This decomposition by deflagration is performed as follows :—One part, by weight, of the insoluble powder is intimately mixed with two parts of dry carbonate of sodium, two parts of fine and pure charcoal powder, and twelve parts of powdered nitre. The mixture is put in a thin porcelain dish, or clean iron tray ; the dish, or little tray, is placed under a hood, or in the open air, and a lighted match is applied to the centre of the heap. The deflagration is completed in two or three seconds, and a well-fused mass remains. This mass is detached from the cooled dish or tray, and boiled with water in a beaker ; it is generally very porous, and is therefore readily disintegrated by stirring it in the hot water with a glass rod. The soluble portion will all be extracted in a very few minutes. The residue left by water is treated with acid precisely as described in § 86. The aqueous and acid solutions of the deflagrated substance are submitted to the same operations as the corresponding solutions of substances fused in crucibles (§ 86). A little charcoal is generally left undissolved by the acid, and with it any of the substance which may have escaped decomposition. The mixture of one part, by weight, of powdered charcoal, and six parts of nitre, may be kept ready mixed for effecting the fusion of insoluble substances.

The advantages of this process are that it is quick, requires only cheap and common tools, and may be applied to substances containing reducible metals, as well as to any others. It is, of course, inapplicable when sodium and potassium are to be sought for in silicates. Chrome-iron-ore cannot be decomposed in this way. The insoluble sulphates, chloride of silver, binoxide of tin, fluorspar, glass, and many natural silicates, may be very well treated by this method, in spite of its apparent roughness.

CHAPTER XII.

TREATMENT OF A PURE METAL OR ALLOY.

91. The elements which are now used in the arts in the metallic state, and which therefore may come into the hands of the analyst as metals, either pure or alloyed, are silver, lead, mercury, bismuth, cadmium, copper, arsenic, antimony, tin, gold, platinum, aluminum, iron, zinc, nickel, and magnesium. These metals can all be brought into solution and detected in the wet way with ease and certainty. It is therefore not worth while to submit a metal, or metallic alloy, to preliminary blowpipe tests, although at need mercury and arsenic can be readily detected by the closed-tube test (§ 76), and many others, by exposing them on charcoal to the reducing and oxidizing flame (compare § 77, p. 106).

A portion of the metal or alloy to be examined should first be reduced to as fine a state of division as possible. If it is brittle, it can be powdered; if soft, shavings can be cut from it; if tough and hard, it can perhaps be fused, and shaken into powder while melted, or granulated by being poured from a height into cold water. Filings should be the last resort, because of the possibility of foreign admixture of iron.

92. *Action of Nitric Acid on the Metals.* A small quantity of the divided metal or alloy, about the equivalent of a pea in bulk, is placed in a flask, covered with concentrated nitric acid, and heated gently under a hood or in the open air for half an hour.

If complete solution ensues, gold, platinum, tin, and antimony are probably altogether absent; they can only be present in very minute proportion. Any of the other metals above enumerated may be present. Transfer the acid solution to a porcelain dish, and evaporate it almost to dryness; dilute the evaporated liquid with five or six times its bulk of water, and proceed with the analysis in the usual way (§ 43). If the solution becomes turbid on the addition of water, bismuth is doubtless present. In this case enough acid must be restored to the solution to clarify it. Mercury, if present, will be dissolved to mercuric nitrate.

If a residue remains undissolved, add a little more acid, to make sure that the acid is incapable of further action; and when this point is settled, test a drop or two of the clear liquid on platinum foil, to ascertain if anything has entered into solution. If the nitric acid has effected a partial solution of the original metal, evaporate the liquid nearly to dryness, dilute the evaporated mixture with water, filter, and submit the filtrate to the usual course of analysis. The residue is thoroughly washed with water, to prepare it for further treatment. Three different cases may occur, readily distinguishable by the mere appearance of the residue.

a. The insoluble substance is non-metallic and white. In this case tin and antimony may be present, but gold and platinum are probably absent. The white residue may contain the insoluble oxides of tin and antimony, or either of them. These elements are to be detected by the methods of § 24, or by the method described just below (first part of *c*).

b. The insoluble substance is metallic, as evidenced by its lustre, if it is in visible fragments, or by the weight and gray or black color of its powder, if it is in a fine state of division. Such a residue must be either gold or platinum (or some of the rare platinum-like metals which lie without the range of this manual). The residue is dissolved in aqua-regia, and evaporated to a very small bulk.

Test for Gold. A portion of this evaporated liquid is diluted with ten times its bulk of water, and poured into a beaker which is placed on a sheet of white paper. A small quantity of a solution of protochloride of tin (§ 143) is tinged yellow by the addition of a few drops of solution of sesquichloride of iron (§ 140), and then considerably diluted. A glass rod is dipped, first into this tin solution, and then into the solution to be tested for gold. If even a trace of the precious metal be present, a blue or purple streak will be observed in the track of the rod. If the quantity of gold be more considerable, a pink tinge will be im-

Test for **Au.**

parted to the solution, or a purplish precipitate will be produced by a sufficient quantity of the tin-solution. This "purple-of-Cassius" test is applicable to very acid solutions.

Test for Platinum. To another undiluted portion of the cooled aqua-regia solution, a cold concentrated solution of chloride of ammonium is added. The formation of a yellow, crystalline precipitate of chloroplatinate of ammonium indicates the presence of platinum (or of some rare platinum-like metal). By adding a little alcohol to the liquid, the test is made more delicate. In a difficult case, the aqua-regia solution might be evaporated to dryness with chloride of ammonium, and the residue treated with weak alcohol and water, which would dissolve all the ingredients except the chloroplatinate. Upon ignition chloroplatinate of ammonium leaves spongy platinum behind.

Test for Pt.

c. The insoluble residue contains both a white powder and a metallic substance. It must then be examined for antimony, tin, gold, and platinum. The following directions presuppose the presence of all four metals—a very rare case.

The residue is placed in a porcelain dish in contact with a slip of clean, smooth, platinum foil, and heated with a little chlorhydric acid. Water is then added, and a small fragment of zinc is put into the liquid. Tin and antimony will be reduced to the metallic state by the zinc, and are detected by the method of § 24, p. 41. Gold and platinum have been from the first in the metallic state, and do not interfere with the processes for detecting tin and antimony. When the antimony has been extracted by tartaric acid, the platinum foil is taken out of the porcelain dish, and the metallic residue from the tartaric acid solution is thoroughly washed by decantation, dissolved in aqua-regia, and tested for gold and platinum, as just described in § 92 *b*.

CHAPTER XIII.

PRELIMINARY EXAMINATION OF A LIQUID.

93. *Evaporation Test.* The first step in the examination of an unknown liquid is to evaporate a few drops at a gentle heat on platinum foil. Attention should be paid to the smell of the escaping vapors in order to ascertain if the solvent be water or some other fluid, like alcohol, ether, benzine, or a strong acid. If no appreciable residue remain, the fluid is probably pure water, or some other volatile liquid ; or it is possible that the liquid is some very dilute solution, like a spring water, which needs extreme concentration before the solid substances dissolved in it can be detected. When a residue remains on the foil, the heat is increased, first, to ascertain if the dissolved substances are wholly volatile, in which case only compounds of ammonium, mercury, arsenic and antimony, can be present ; and, secondly, to ascertain if there be any organic matter in the liquid. Carbonization or charring, with the attendant phenomena (§ 76, I), occurs when fixed organic matter is present. If organic matter is discovered, it must be destroyed by the second method of § 76, I, before the analysis can be proceeded with. A volatile organic solvent can, of course, be got rid of by a simple evaporation to dryness.

94. *Testing with Litmus.* The next step is to test the solution with litmus-paper.

a. If it is neutral, and the solvent is water, consult § 80.

b. If it is acid, the acidity may be due to a normal salt having an acid reaction, or to an acid salt, or to free acid. No general inferences can be drawn from the acid reaction, except that carbonates and sulphides are absent. If dilution of the acid fluid produces turbidity, the presence of antimony or bismuth may be inferred.

c. If it is alkaline, consult § 80 (*Alkaline*).

95. By evaporating a portion of the original solution to dryness, the dissolved solid is obtained. This solid may be subjected to the whole of the preliminary treatment prescribed for a salt, mineral, or other non-metallic solid (§§ 76, 77); but inasmuch as the main object of all preliminary treatment of a solid is to learn how to get it into solution with the least difficulty, it is seldom worth while for the analyst to make a solid out of a solution, and thus forego the advantage of having the solution already made to his hand.

96. *Testing for Ammonia.* A small portion of the original solution must always be tested for ammonium salts, by heating it in a test-tube with an equal bulk of slaked lime. The gas is recognized by its smell and its reaction with chlorhydric acid (§ 76).

The means of identifying and isolating the rare elements, the methods by which minute traces of one substance may be detected when hidden in proportionally large quantities of other substances, as when the impurities of chemicals and drugs are exhibited, and the processes to be employed in special cases of peculiar difficulty, such as the analysis of complex insoluble minerals, or the detection of mineral poisons in masses of organic matter, must be studied in complete treatises upon chemical analysis, or in works specially devoted to these technical matters. Such details, however valuable to the professional analyst, or expert, would not be in harmony with the plan of this manual.

PART THIRD.

CHAPTER XIV.

REAGENTS.

[** Those reagents in the following list which are marked with the double asterisk are rarely employed; a single small bottle of each of them in the laboratory will be enough for many students.]

97. *Chlorhydric Acid (Concentrated).* The strong common acid prepared by chemical manufacturers, though usually far from pure, will answer for most of the purposes of this manual. It must, however, be continually borne in mind that the commercial acid is usually contaminated with sulphuric acid, and very often with traces of arsenic and iron. These impurities may be present in sufficient quantity to render the acid unfit for use when these very substances are to be tested for in the mixture to be analyzed.

The yellow color of the commercial acid, though often attributed to iron, is really due for the most part to the presence of a peculiar organic compound which is soluble in the strong acid.

Pure acid may be prepared by distilling a mixture of fused chloride of sodium and sulphuric acid, and collecting the gas in water (see the authors' Manual of Inorganic Chemistry, Exp. 49).

If in any experiment doubts arise concerning the character of a reagent, a quantity of it, somewhat larger than that which has been mixed with the substance under examination, should be tested by itself, and the reaction compared with that exhibited in the doubtful case. If the result of this trial is unsatisfactory, the experiment must be repeated with reagents which are known to be pure.

98. *Chlorhydric Acid* (*Dilute*). Mix 1 volume of the common concentrated acid, or—where special purity is required—of the pure strong acid, with 4 volumes of water.

99. *Nitric Acid* (*Concentrated*). Use the colorless commercial acid of 1.38 or 1.40 specific gravity. Strong nitric acid of tolerable purity can usually be obtained from the dealers in coarse chemicals. An acid, which when diluted with five parts of water gives no decided cloudiness with either nitrate of silver or nitrate of barium, is good enough for most uses in qualitative analysis.

100. *Nitric Acid* (*Dilute*). Mix 1 volume of the strong acid with 5 volumes of water.

101. *Aqua-regia* should be prepared only in small quantities, at the moment of use, by mixing in a test-tube one volume of *strong* nitric acid, with three or four times as much *strong* chlorhydric acid.

102. *Sulphuric Acid* (*Concentrated*). The oil of vitriol of commerce will usually be found pure enough for the purposes of this manual.

103. *Sulphuric Acid* (*Dilute*) is prepared by gradually adding 1 part of the concentrated acid to 4 parts of water contained in a beaker or porcelain dish; the mixture must be constantly stirred with a glass rod. When the mixing is finished, the liquid is left at rest until all the sulphate of lead which has separated from the strong acid has settled to the bottom; the clear liquid is then decanted into bottles.

104. *Oxalic Acid.* Dissolve 1 part, by weight, of the commercial crystals in 20 parts of water.

105. *Acetic Acid.* The ordinary commercial acid.

106. *Tartaric Acid* should be kept in the state of powder, since solutions of it slowly decompose. For use, dissolve a small portion of the powder in two or three times its volume of hot water.

107. *Sulphuretted Hydrogen Gas* (*Sulphydric Acid*) is prepared, as needed, by acting upon fragments of sulphide of iron with dilute sulphuric acid in the apparatus described in §§ 178, 179. The apparatus should always be placed either in the open air, or in a strong draught beneath a chimney.

108. *Sulphuretted Hydrogen Water.* Pass sulphuretted hydrogen gas into a bottle of water until the water can absorb no more. To determine when the absorption is complete, close the mouth of the bottle tightly with the thumb, and shake the liquid. If the water is saturated, a small portion of the gas will be set free by the agitation, and a slight outward pressure against the thumb will be felt. If the water is not fully saturated, the agitation will enable it to absorb the gas which lay in the upper part of the bottle, and a partial vacuum will be created, so that an inward pressure will be felt.

Since sulphuretted hydrogen water soon decomposes when exposed to the air, it should always be kept in tightly closed bottles; and no very large quantity of it should be prepared at once. A good way of keeping the solution is, to fill a number of small phials with the fresh liquid, cork them tightly, and invert them in water, so that their necks shall always be immersed and protected from the atmosphere.

At the moment of using this reagent, its quality should always be proved by smelling of it, or by adding a drop or two of the liquid to a drop of acetate of lead.

109. *Ammonia-Water.* Commercial aqua-ammoniæ may usually be obtained pure enough for the purposes of this manual. Dilute 1 volume of the strong liquor with 3 volumes of water. Ammonia-water should be free from carbonic acid; when diluted, as above, it ought not to yield any precipitate when tested with lime-water.

110. *Sulphydrate of Ammonium.* Pass sulphuretted hydrogen gas through ammonia-water, diluted as described in § 109, until a portion of the liquid yields no precipitate when tested with a drop of a solution of sulphate of magnesium.

Since sulphydrate of ammonium decomposes after a while, when exposed to the air, it is not advisable to prepare it in large quantities. In case any doubt arises as to the quality of the reagent, add some of it to a drop of acetate of lead. Unless a dense, black precipitate of sulphide of lead is immediately thrown down, the sulphydrate is worthless.

111. *Carbonate of Ammonium.* Dissolve 1 part, by weight, of the commercial salt in 4 parts of water, and add to the mixture 1 part of strong ammonia-water.

112. *Chloride of Ammonium.* Dissolve 1 part, by weight, of the crystallized commercial salt in 10 parts of water.

113. *Oxalate of Ammonium.* Dissolve 1 part, by weight, of the salt in 24 parts of water.

114. *Nitrate of Ammonium.* The commercial salt, kept as dry as possible, in the form of small crystals.

115. ** *Molybdate of Ammonium.* Digest 1 part, by weight, of molybdic acid for some hours in 4 or 5 parts of strong ammonia-water, and mix the clear solution with 12 or 15 parts of strong nitric acid ; or dissolve 1 part of molybdate of ammonium in 3 or 4 parts of weak ammonia-water, and mix the liquid with 12 or 15 parts of nitric acid, as before.

116. *Caustic Soda.* Place 1 part, by weight, of the best commercial caustic soda in a large stoppered bottle ; pour upon it 8 or 9 parts of water, and shake the bottle at intervals, until the whole of the soda has dissolved. Leave the bottle at rest until the liquid has become clear, and finally transfer the solution, with a siphon, to the small bottles in which it is to be kept for use. The solution thus prepared, though pure enough for the uses prescribed in this manual, is really far from pure. It would be unfit for use in a delicate research, because it is usually contaminated with chloride, sulphate and carbonate of sodium, and is liable to contain traces of aluminate, phosphate, and silicate of sodium. Since some nitrate of sodium is added to it in the process of manufacture, the soda is liable to be contaminated with this salt and the products of its decomposition, including ammonia. This last impurity is liable to be given off when the solution is boiled.

Caustic potash, as prepared for surgeons' use, may be substituted for caustic soda whenever it can be more readily obtained. The potash should be dissolved in about 10 parts of water.

Since solutions of the caustic alkalies act upon glass rather easily, especially when its outer surface or "fire-glaze" has once been removed, it often happens, when the soda solution is kept in glass-stoppered bottles, that the stoppers become immovably cemented to the glass by the silicate of sodium which forms in their necks. This difficulty may be avoided by wiping the necks of the bottles dry after any of the solution has been poured from them; but it will usually be found more convenient to replace the glass stoppers with plugs of vulcanized caoutchouc, or, better still, with small glass stoppers, over the bodies of which short pieces of caoutchouc tubing have been stretched.

117. *Sulphydrate of Sodium.* Dissolve 1 part, by weight, of commercial sulphide of sodium in 8 parts of water. Sulphide of potassium is not an available substitute for sulphide of sodium.

118. *Carbonate of Sodium.* The anhydrous salt of commerce will answer for most uses. It should not contain much sulphate of sodium. For those cases in which the use of carbonate of sodium, free from any contamination of sulphate, is prescribed, the salt may be prepared by washing a pound or two of bicarbonate of sodium repeatedly upon a filter, with small quantities of ice-cold water, until the original quantity is reduced to a fifth or a sixth of its bulk.

119. *Biborate of Sodium.* Common borax, powdered.

120. *Nitrate of Sodium.* Select a white, clean sample of the commercial salt, and dissolve 1 part, by weight, in 3 parts of water.

121. *Diphosphate of Sodium.* Dissolve 1 part, by weight, of "common phosphate of soda" in 10 parts of water.

122. ** *Acid Sulphate of Sodium.* Heat a mixture of 16 parts, by weight, of Glauber's salt, and 5 parts of concentrated sulphuric acid, in a platinum vessel, until a portion of the melted mass becomes distinctly solid when taken up on a glass rod. Then allow the mixture to become cold; remove the cold lump from the platinum vessel, and break it into fragments. Keep the coarse powder in a tight, glass-stoppered bottle.

123. *Sulphate of Potassium.* Dissolve one part, by weight, of the crystallized salt in 200 parts of water. A solution of this strength contains the same proportional quantity of sulphuric acid as is contained in a saturated aqueous solution of sulphate of calcium. Hence it cannot precipitate the latter when added to solutions of the soluble calcium salts.

124. *Chromate of Potassium.* (The normal or "neutral" yellow chromate.) Dissolve 1 part, by weight, of the salt in 8 parts of water.

125. *Ferrocyanide of Potassium.* (*Yellow Prussiate of Potash.*) Dissolve 1 part, by weight, of the commercial salt in 12 parts of water.

126. ** *Ferricyanide of Potassium.* (*Red Prussiate of Potash.*) Since the aqueous solution of this salt undergoes decomposition, with formation of some ferrocyanide, when kept for any length of time, the salt should be kept for use in the form of powder. The commercial salt is pure enough for analytical purposes. A minute fragment of it may be dissolved in water at the moment of use.

127. ** *Cyanide of Potassium.* The better sorts of the commercial article are pure enough for analytical purposes. It should be kept in the solid form, in a tightly stoppered bottle. When the solution is required, dissolve 1 part of the salt in 4 parts of cold water.

128. *Nitrate of Potassium.* Refined saltpetre may be employed. It should be kept in the state of powder.

129. ** *Nitrite of Potassium.* Weigh out 8 parts of concentrated nitric acid, mix it with an equal weight of water, and place the mixture in a glass flask provided with a perforated cork and gas delivery-tube. The flask should be so large that the mixture only half fills it. Throw into the liquid two parts of starch, in lumps, and heat the mixture until red fumes of nitrous and hyponitric acids begin to be given off; then remove the lamp lest the action become too violent. Conduct the fumes into a bottle containing 5 parts of potash-lye of 1.27 sp. gr., until the latter is saturated. Then filter the saturated liquid, and evaporate it to dryness. For use, dissolve 1 part of the dry salt in 2 parts of water.

The nitrite of potassium bought of dealers in fine chemicals is often unfit for the uses prescribed in this manual; it can readily be made good, however, by dissolving it in water, in the proportion above given, and saturating the solution with nitrous fumes.

130. ** *Iodide of Potassium and Starch Papers.* Dissolve a gramme of pure iodide of potassium (free from iodate) in 200 cubic centimetres of water. Heat the solution moderately in a porcelain dish, and stir into it ten grammes of starch which has been reduced to the consistence of cream by rubbing it in a mortar with a small quantity of water. Stir the mixture until it gelatinizes, taking care not to burn the starch, then allow the paste to cool, and spread it thinly upon one side of white glazed paper with a wooden spatula. Dry the paper, cut it into strips as large as the little finger and preserve it in stoppered bottles kept carefully closed.

131. *Nitrate of Silver.* Dissolve 1 part, by weight, of the commercial crystals in 20 parts of water.

132. *Slaked Lime.* Mix common quicklime with half its weight of water. Keep the powder in bottles with tight stoppers.

133. ** *Lime Water.* Place a handful of slaked lime in a large bottle, pour in enough water to almost fill the bottle, cork the latter tightly, and shake it at intervals during several days. Decant the clear liquid into smaller bottles for use. Refill the large supply-bottle with water, and again shake it at intervals.

134. *Chloride of Calcium.* Stir powdered white marble into dilute chlorhydric acid until the acid is saturated, and dilute 1 part of the concentrated solution with 5 parts of water.

135. *Chloride of Barium.* Dissolve 1 part, by weight, of the commercial salt in 10 parts of water.

136. ** *Nitrate of Barium.* Dissolve 1 part, by weight, of the commercial salt in 15 parts of water.

137. *Acetate of Lead.* Dissolve 1 part, by weight, of "sugar of lead" in 10 parts of water.

138. *Lead-paper.* Wet strips of white paper in a solution of acetate of lead, or better, in a solution of subacetate of lead, and dry them in air which is free from sulphuretted hydrogen. Cut the dried paper into slips as large as the little finger, and keep the slips in tightly stoppered bottles. Or, paper may be slightly moistened with a solution of acetate of lead at the moment of use.

139. *Sulphate of Magnesium and Chloride of Ammonium.* Dissolve 24.6 grammes of Epsom salt and 33 grammes of commercial chloride of ammonium in water, add some ammonia-water to the solution and dilute the liquor to the volume of a litre. If less than a litre of the reagent is required, the weights above given may, of course, be reduced in any desired proportion. Filter the solution to separate any precipitate of ferric hydrate or other insoluble matters which may have been present as impurities in the components of the mixture, and preserve the clear liquid.

From a solution thus prepared no hydrate of magnesium can be precipitated by ammonia-water ; herein consists the advantage of the mixture as a test for phosphoric and arsenic acids.

140. ** *Ferric Chloride.* This solution is prepared by passing chlorine gas through a saturated solution of iron tacks in chlorhydric acid, until a drop of the fluid no longer produces a blue precipitate in a solution of ferricyanide of potassium. The solution is then heated to expel the excess of chlorine.

141. ** *Nitrate of Cobalt.* Dissolve 1 part, by weight, of the crystallized salt in 10 parts of water.

142. ** *Sulphate of Copper.* Dissolve 1 part, by weight, of the crystallized salt (blue vitriol) in 10 parts of water.

143. ** *Protochloride of Tin.* This solution is prepared by boiling scraps of tin with strong chlorhydric acid until hydrogen ceases to be evolved. The tin must be in excess. The solution is diluted with four times its bulk of water acidulated with chlorhydric acid, and filtered, if necessary. The clear liquid must be kept in a tightly closed bottle containing some bits of tin.

144. ** *Red Oxide of Mercury.* The commercial oxide. It should leave no residue when heated upon platinum foil.

145. ** *Chloride of Mercury.* Dissolve 1 part of "corrosive sublimate" in 16 parts of water.

146. ** *Bichloride of Platinum.* Cut a small quantity of worn-out platinum foil into very fine pieces and boil them in a porcelain dish, with successive small portions of aqua-regia (§ 101), until all the metal has been dissolved. Collect the several portions of aqua-regia, partially saturated with platinum, in another dish, and evaporate the liquid to dryness on a water bath. Dissolve the residue in 10 parts of water for use.

147. ** "*Solution of Indigo*" (*Sulphindigotic Acid*). Pour 5 parts (5 grammes will be ample) of fuming sulphuric acid into a beaker, place the latter in a dish of water to keep it cool, and stir into the acid, little by little, 1 part of finely powdered indigo. When all the indigo has been added to the acid, leave the mixture at rest for 48 hours; then pour it into 20 times its own volume of water, filter the mixture, and preserve the filtrate for use.

148. *Litmus Paper.* Heat 1 part, by weight, of commercial litmus with 6 parts of water, upon a water bath for several hours, taking care to replace the water which evaporates. Filter, divide the filtrate into two equal portions, and stir one-half repeatedly with a glass rod dipped in very dilute nitric acid, until the color appears distinctly red. Pour the blue and red halves into a porcelain dish, and stir the mixture. Draw strips of fine unsized paper through the liquid, and hang them on cords to dry. The color of the paper thus obtained is not blue, but bluish-violet. It turns blue when touched with an alkali, and red when exposed to acids, and may be used indifferently as a test for either acids or alkalies.

149. *Turmeric Paper.* Digest 1 part, by weight, of bruised turmeric in 6 parts of warm spirits of wine. Filter the tincture and steep in it narrow strips of white paper. The dried paper should have a fine yellow color.

150. *Starch Paste* should be prepared, when wanted for use, by boiling 30 cubic centimetres of water in a porcelain dish, and stirring into it half a gramme of starch which has previously been reduced to the consistence of cream by rubbing it in a mortar with a few drops of water.

151. *Water.* Clean rain-water will serve well enough for most of the purposes of this manual. In granitic regions the water of many lakes, brooks, and ponds also, is nearly pure. Pure water may be obtained by melting blocks of compact ice, or by distilling ordinary water in glass or copper retorts and rejecting the first portions of the distillate. It should yield no precipitate when tested with chloride of barium and nitrate of silver.

CHAPTER XV.

UTENSILS.

152. The implements required by the student of qualitative analysis are few and simple. Besides bottles for the reagents enumerated in the foregoing list, and a few small phials for the preservation of samples of salts and mixtures to be analyzed, there will be needed :—

A dozen test-tubes,
A wooden test-tube rack,
A test-tube brush,
A nest of small beakers,
2 or 3 glass stirring-rods,
3 small glass funnels,
1 small glass flask,
1 wash-bottle,
A gas-bottle for generating sulphuretted hydrogen,
A common jewellers' blow-pipe,
A pair of small iron pincers (jewellers' tweezers),
A piece of platinum foil,
A bit of platinum wire,

2 small evaporating dishes,
1 porcelain crucible,
1 triangle of iron wire,
An iron ring-stand,
A filter-stand,
A lamp,
A few packages of cut filters or a quire of filter paper,
A few corks or caoutchouc stoppers,
A piece of blue cobalt glass (see § 41).

153. *Reagent Bottles.* The bottles in which reagents are kept should be of cylindrical shape, and rather high than wide. They should be closed with glass stoppers which fit accurately, but are not very finely ground. The stoppers should have upright (not mushroom-shaped) heads. Most of the liquid reagents may be conveniently kept in narrow-mouthed bottles of the capacity of 6 fluid ounces; but to avoid the necessity of frequently refilling the bottles, it is well to keep the solutions most commonly employed—namely, dilute chlorhydric and nitric acids, caustic soda, ammonia-water, chloride of ammonium, and carbonate of ammonium—in 8-ounce bottles. Care must be taken in this case to choose bottles of such shape that they can be readily grasped between the thumb and fingers.

For the reagents which are to be kept in the dry state, wide-mouthed bottles of the capacity of 2 or 3 ounces should be chosen.

Reagent bottles should always be made "extra-heavy," since, from constant use, they are exposed to many blows. The lustrous "flint-glass bottles" of American or English make are ill suited for the preservation of liquid reagents; for such glass is easily attacked by many chemical agents, and is therefore likely to render the reagents impure. German bottles are usually to be preferred. They are made of glass free from lead, have round shoulders and well-ground stoppers, and are often numbered both upon the bottle and the stopper, so that the proper place of the latter can always be discovered. French bottles, though made of good glass and sold at a low price, have usually such square shoulders that it is well nigh impossible to empty them completely, and often difficult to pour out a liquid from them drop by drop. Their stoppers, moreover, are usually too finely ground, and are hence constantly liable to stick fast.

Each reagent bottle should be kept in a particular place on shelves before the operator and convenient to his hand. Whenever a reagent is to be used, the bottle which contains it should be grasped in the right hand; the stopper should be taken out by pinching it between the first and second, or third and fourth, fingers of the left hand, or by pressing it between the little finger and palm of that hand. In either case, the bottle is withdrawn from the stopper, and not the stopper from the bottle. Neither bottle nor stopper should be put upon the table; the stopper should be held in the left hand as long as the bottle is open. When the reagent has been poured out, the bottle is immediately closed, and returned to its own place upon the shelf. If these apparently trifling particulars are scrupulously attended to, no stopper can ever be misplaced, or soiled by contact with liquids or dirt on the table; and the bottle will always be found in its proper place when instinctively reached for. Moreover, the label on the bottle cannot be injured by drops of the reagent, since the liquid must necessarily be poured from the back, or blank, side of the bottle.

154. When a stopper sticks tightly in the neck of a bottle, it may sometimes be loosened by pressing it first upon one side, and then upon the other, with the thumb of the right hand, while the fingers of that hand grip the bottle, and the bottle is held still with the left hand. Or the neck of the bottle may be immersed in hot water for a minute or two, to expand the glass outside the stopper. The stopper can then usually be taken out without trouble. The hot water may be conveniently applied by pouring a slow stream of it from a wash-bottle upon the neck of the bottle. Another way is to heat the neck of the bottle over a very small flame of the gas or alcohol lamp. No matter how the glass is heated, the bottle must be constantly turned round and round, in order that each side of the neck may be equally exposed to the heat, and the risk of cracking the bottle so be lessened.

155. *Test-tubes* are little cylinders of thin glass with round thin bottoms and lips slightly flared. Their length may be from five to seven inches, and their diameter from one-half to three-fourths of an inch ; they should never be so wide that the open end cannot be closed by the ball of the thumb.

Test-tubes are used for heating small quantities of liquid over the gas or spirit-lamp ; they may generally be held by the upper end in the fingers without inconvenience ; but in case they become too hot to be held in this way, a strip of thick folded paper may be nipped round the tube, and grasped between the thumb and forefinger just outside the tube.

Two precautions are invariably to be observed in heating test-tubes :—1st. The outside of the tube must be wiped perfectly dry ; and 2nd. The tube must be moved in and out of the flame for a minute or two when first heated. It should be rolled to and fro also to a slight extent between the thumb and forefinger, in order that each side of it may be equally exposed to the flame. A drop of water on the outside of the tube keeps one spot cooler than the rest. The tube breaks, because its parts, being unequally heated, expand unequally, and tear apart.

When a liquid is boiling actively in a test-tube, it sometimes happens that portions of the hot liquid are projected out of the tube with some force ; the tube should therefore always be held in an inclined position, and the operator should be careful not to direct it towards himself, or towards any other person in his neighborhood.

Test-tubes are cleaned by the aid of cylindrical brushes, made of bristles caught between twisted wires, like those used for cleaning lamp-chimneys ; the brushes should have a round end of bristles.

156. *Test-tube Rack.* Test-tubes are kept in a wooden rack, such as is represented in Figure 1. When in use, the tubes stand upright in the holes of the rack ; but clean tubes are inverted upon the pegs behind the holes, in order that they may be kept free from dust, and that the last portions of wash-water may drain away from them after washing. The rack should be large enough to hold a dozen tubes.

Fig. 1.

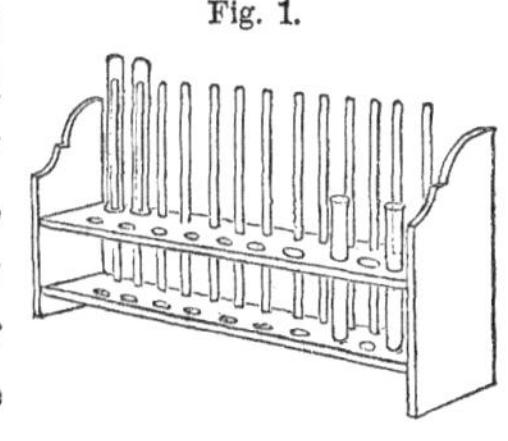

157. *Flasks.* Small Berlin flasks, of two or three ounces capacity, are well suited for the purposes of qualitative analysis. These German flasks are tough, capable of withstanding sudden changes of temperature, and durable, although their bottoms and sides have all the requisite thinness. When a liquid is to be boiled in a flask, the flask should be placed upon a support of wire-gauze (§ 165), and sufficiently inclined, to prevent any particles of the liquid from being thrown out of the neck by the movement of ebullition.

As with test-tubes, and all other glass or porcelain vessels, of whatever form, the outside of a flask must be wiped perfectly dry before exposing it to the lamp. The flame should be moved about also beneath the flask, at first, in order that the temperature of the latter may be raised equally and not too rapidly.

158. *Beakers* are thin, flat-bottomed tumblers, with a slightly flaring rim. They are bought in sets or nests. A nest in which the largest-sized beaker has a capacity of about 6 ounces will be sufficient for the requirements of this work.

159. *Glass Funnels* should be thin and light, and should be about 2 or 2.5 inches in diameter. Their sides should incline at an angle of 60°. The wider the throat of the funnel the better.

160. *Filtering.* Paper filters are employed in qualitative analysis to separate precipitates from the liquids in which they have been formed. A good filtering-paper must be porous enough to filter rapidly, and yet sufficiently close in texture to retain the finest powders ; and it must also be strong enough to bear, when wet, the pressure of the liquid which is poured upon it. Filter-paper should never contain any gypsum or other soluble material, and should leave only a small proportion of ash when burned. White or light-gray paper is to be preferred to colored, since it more commonly fulfils these requirements.

Filtering-paper is commonly sold in sheets, which may be cut into circles of any desired diameters for use, according to the various scales of operation and quantities of liquids to be filtered. Or packages of "cut-filters" may be procured ready-made from the dealers in chemical wares.

As a general rule, small filters should be employed in analytical operations ; the mixture to be filtered should be poured, by small successive portions, upon a filter no larger than is needed to hold the whole of the solid matter which is to be collected. Filters about three inches in diameter are well suited for most of the analytical operations described in this work, though there are many cases where smaller filters are required, and a few instances in which filters as large as four inches in diameter might be necessary.

There are two ready methods of preparing filters for use. According to the first method, shown in Fig. 2, a circle of paper is folded over on its own diameter, and the semicircle thus obtained is folded once upon itself into the form of a quadrant ; the paper thus folded is opened so that three thicknesses shall come upon one side, and one thickness upon the other, as shown in the upper half of Fig. 2 ; the filter is then placed in a glass funnel, the angle of which should be precisely that of the opened paper, viz., 60°. The paper may be so folded as to fit a funnel whose angle is more or less than 60° ; but this is the most advantageous angle, and funnels should be selected with reference to their correctness in this respect.

Fig. 2.

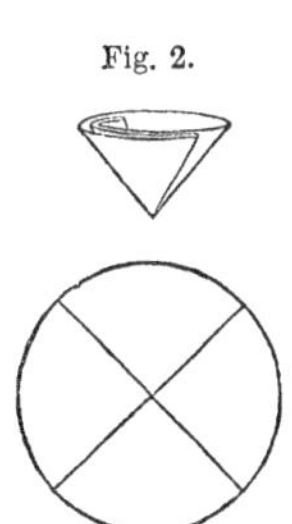

Fig. 3.

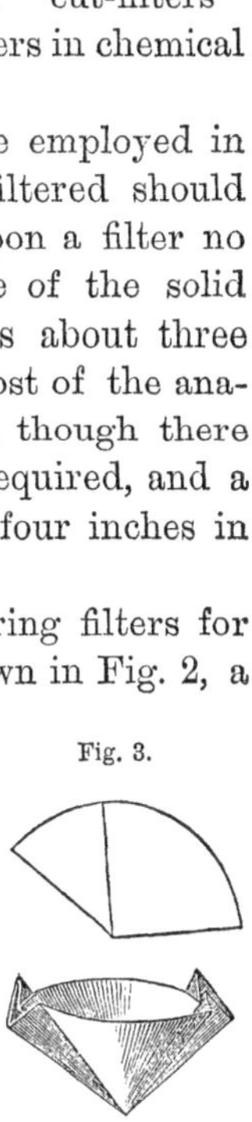

In the second method of folding filters, the circle of

paper is doubled once upon itself, as before, into the form of a semicircle, and a fold equal to one quarter of this semicircle is turned down on each side of the paper. Each of the quarter semicircles is then folded back upon itself, as shown in the lower half of Fig. 3 ; the filter is opened, without disturbing the folded portions, and placed in the funnel. Filtration can be rapidly effected with this kind of filters, for the projecting folds keep open passages between the filter and the funnel, and thus facilitate the passage of the liquid. That portion of the circle of paper, which must necessarily be folded up in order to give the requisite conical form to a paper filter, retards filtration in the first manner of folding, but helps it in the second.

A filter should always be moistened with water after it has been placed in the funnel, in order that the fibres of the paper may be swollen and the size of its pores diminished, before any of the matter to be filtered can pass into them.

Coarse and rapid filtration—as in the preparation of reagents—can be effected with paper filters of large size, or with cloth bags ; also by plugging the neck of a funnel or leg of a siphon loosely with tow or cotton. If a very acid or very caustic liquid, which would destroy paper, cotton, tow, or wool, is to be filtered, the best substances wherewith to plug the neck of the funnel are asbestos and gun-cotton, neither of which is attacked by such corrosive liquids.

161. *Filter-Stand.* Filters less than two inches in diameter may be placed directly in the mouth of a test-tube without need of even a funnel to support them ; and in general the funnel which holds a filter may be thrust directly into the mouth of a test-tube whenever the quantity of liquid to be filtered is small, if only an ample exit be provided for the air in the tube, in the manner shown with the bottle of Fig. 4.

Fig. 4.

But when the quantity of liquid to be filtered is comparatively large, or the operations to which the filtrate is to be subjected require that it should be collected in a beaker or porcelain dish, the funnel should have an independent support. The iron ring-stand, described in § 165, may be used for this purpose in case of need; but wooden stands of the form represented in Fig. 5, adapted expressly for holding funnels, are very convenient and not expensive. The horizontal bar which holds the funnel may be fixed at any height on the vertical square rod by means of a wedged-shaped key, whose form is shown in the figure. A fine-grained wood, which does not swell or shrink much, is desirable for filter-stands.

Fig. 5.

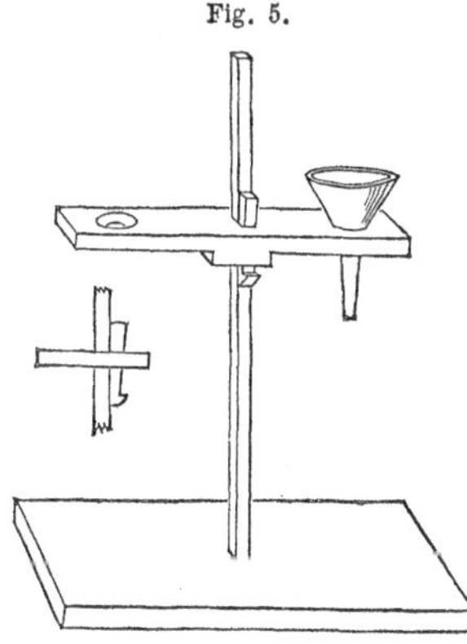

In general, care should be taken that the lower end of the funnel touch the side or edge of the vessel into which the filtrate descends, in order that the liquid may not fall in drops, but run quietly without splashing.

162. *Porcelain Dishes and Crucibles.* Small open dishes which will bear heat without cracking, are much used for boiling and evaporating liquids. The best evaporating-dishes are those made of

Berlin porcelain, glazed both inside and out, and provided with a little lip projecting beyond the rim. The dishes made of Meissen porcelain are not glazed on the outside, and are not so durable as those of Berlin manufacture ; but they are much cheaper, and with proper care last a long time.

The small Berlin dishes, Nos. "00," "0," and "1," are well suited for all the requirements of this work. They will generally bear an evaporation to dryness and subsequent ignition over the open flame of a gas lamp—as when ammonium salts are expelled from Class VII (§ 41) —but it is well to protect the dish somewhat by placing it upon a piece of wire-gauze, rather than to support it upon a simple triangle. The Meissen dishes do not so well endure an open flame. The cheaper kinds of evaporating dishes, made of "semi-porcelain," should never be subjected to this severe treatment ; they are, for that matter, unfit for use in qualitative analysis.

Very thin, highly glazed porcelain crucibles, with glazed covers, are made both at Berlin and at Meissen near Dresden. In general the Meissen crucibles are thinner than the Berlin, but the Berlin crucibles are somewhat less liable to crack ; both kinds are glazed inside and out, except on the outside of the bottom. The Berlin, Nos. "00" and "0," respectively 1 1-4 and 1 1-2 inches in diameter, are best suited for the purposes of this manual. As the covers are much less liable to be broken than the crucibles, it is advantageous to buy more crucibles than covers, whenever it is possible so to do. Porcelain crucibles are supported over the lamp on an iron wire triangle ; they must always be gradually heated, and never brought suddenly in contact with any cold substance while they are hot.

163. *Lamps.* The common spirit-lamp will be understood without description, from the figure (Fig. 6). When not actually lighted, the wick must be kept covered with the glass cap; for if the wick were exposed to the air, the alcohol in the spirit upon it would evaporate faster than the water, and the cotton would soon become water-soaked and incapable of being lighted.

Fig. 6.

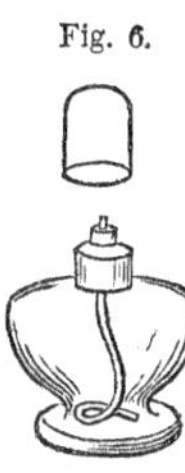

Whenever gas can be obtained, gas-lamps are greatly to be preferred to the best spirit-lamps. For all ordinary uses, the gas-lamp known as Bunsen's burner may be employed. The cheapest and best construction of the lamp may be learned from the following description with the accompanying figures. (Fig. 7.) The single casting of brass *a b* comprises the tube *b* through which the gas enters, and the block *a* from which the gas escapes by two or three fine vertical holes passing through the screw *d*, and issuing from the upper face of *d*, as shown at *e*. The length of the tube *b* is 4.5 c. m., and its outside diameter varies from 0.5 c. m. at the outer end to 1 c. m, at the junction with the block *a*. The outside diameter of the block *a* is 1.6 c. m., and its outside height without the screws is 1.8 c. m. By the screw *c*, the piece *a b* is attached to the iron foot *g*, which may be 6 c. m. in diameter. By the screw *d*, the brass tube *f* is attached to the casting *a b*. The diameter of the face *e*, and therefore the internal diameter of the tube *f*, should be 8 m. m. The

Fig. 7.

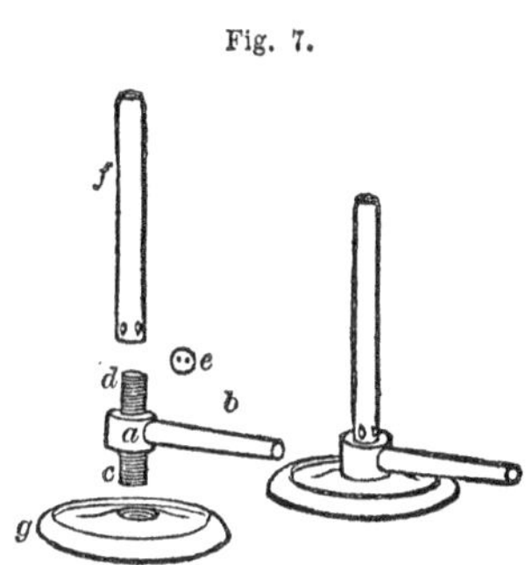

length of the tube *f* is 9 c. m. Through the wall of this tube, four holes 5 m. m. in diameter, are to be cut at such a height that the bottom of each hole will come 1 m. m. above the face *e* when the tube is screwed upon *a b*. These holes are of course opposite each other in pairs. The finished lamp is also shown in the figure. To the tube *b* a caoutchouc tube of 5 to 7 m. m. internal diameter is attached; this flexible tube should be about 1 m. long, and its other extremity should be connected with the gas-cock through the intervention of a short piece of brass gas-pipe screwed into the cock.

In cases where a very small flame is required, as, for example, in evaporating small quantities of liquid, a piece of wire-gauze somewhat larger than the opening of the tube *f* should be laid across the top of the tube, and its projecting edges pressed down tightly against the side of the tube, before the lamp is lighted. In default of this precaution, the flame of a Bunsen's burner, when small and exposed to currents of air, is liable to pass down the tube, and ignite the gas at *d*.

Fig. 8.

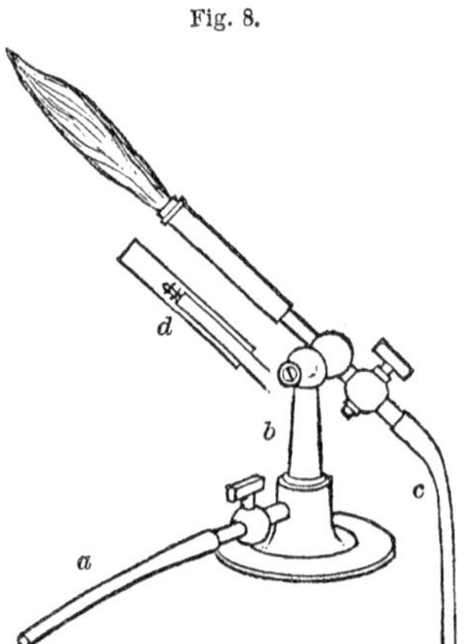

164. *Blast-lamps and Blowers.* Though well suited for all the ordinary operations of the laboratory, the lamps above described are incapable of yielding a very intense heat. Hence, when the contents of a platinum crucible are to be fused or intensely heated, a blast-lamp will be found useful. The best form is that sold under the name of Bunsen's Gas Blowpipe. Its construction and the method of using it may be learned from the accompanying figure ; *a b* is the pipe through which the gas enters, *c* is the tube for the blast of air ; the relation of the air-tube to the external gas tube is shown at *d* ; there is an outer sliding tube by which the form and volume of the flame can be regulated.

If gas is not to be had, a lamp burning oil or naphtha may be employed. Figure 9 represents a common tin, glass-blower's lamp, suitable for burning oil. A large wick is essential, whether oil or naphtha be the combustible.

Fig. 9.

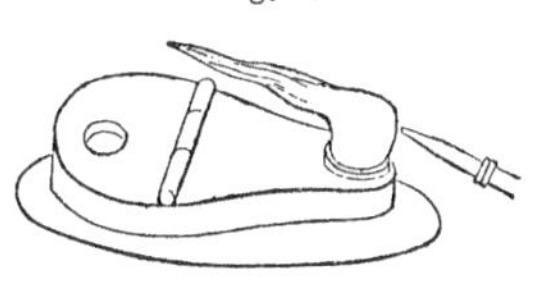

For every blast-lamp a blowing machine of some sort is necessary. To supply a constant blast it is essential that the bellows be of that construction called double. Figures 10 and 11 represent two forms of blowpipe-table ; their height is that of an ordinary table, from which dimension the other proportions may be inferred. A small double-acting bellows is made (formerly used by dentists), which is available at any table by the help of a caoutchouc tube to conduct the blast to the jet. These bellows are too small to give a steady flame of large size, but will, nevertheless, answer for most of the glass-blowing necessary in the execution of the experiments described in this manual.

Fig. 10.

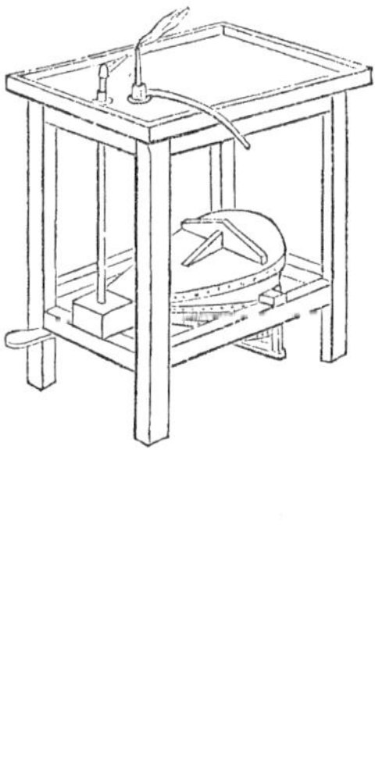

Where an abundant supply of water is at command, the following blowing apparatus is very convenient : A tin pipe *a b* (Fig. 12), about 1 metre long and 13 m. m.

in diameter, has two smaller pipes, 12 to 16 c. m. long, soldered into it ; these small pipes are 8 m. m. in diameter ; one, *c d*, is inserted at right angles 12 c. m. from the end, the other, *e f*, 2.5 c. m. lower, at an angle of 45°. The lower end of the tube passes through the cork of a

Fig. 11. Fig. 12.

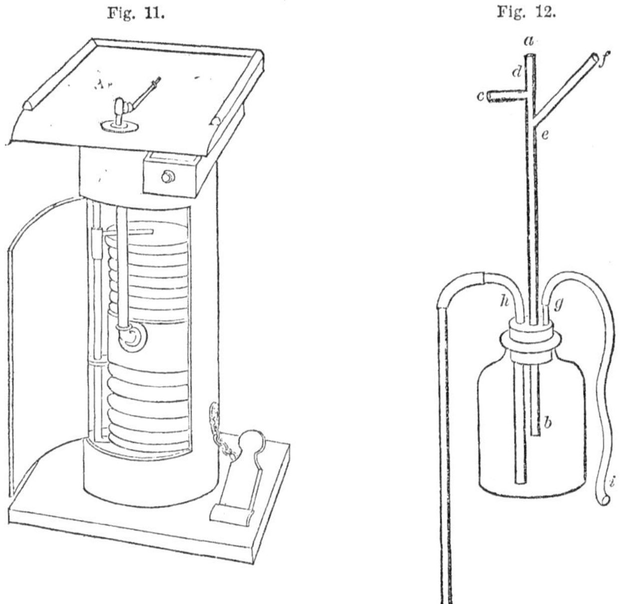

wide-mouthed ten-litre bottle, extending rather more than half way down. Two glass tubes also pass through the cork of the bottle—a short small tube *g*, No. 4, (§ 171) which should reach some 16 c. m. above the cork, but should not project into the bottle, and a larger tube *h*, No. 2, extending to the bottom of the bottle. The outer end of the tube *h* bends over, and is connected by caoutchouc tubing with a straight tube of equal diameter.

This last arrangement forms the siphon. To the tube *g* a caoutchouc tube, *g i*, is attached to convey the blast to any desired point. To produce a blast, the water-cock is connected with the tube *c d* by means of a caoutchouc tube. When the water is turned on, the caoutchouc tube *g i* is closed for a moment with the thumb and finger. This starts the water through the siphon, and immediately a continuous and powerful blast of air rushes out through the tube *g i*, and may be carried directly to the blow-pipe. The siphon must be capable of carrying off a larger stream of water than that which is allowed to enter, so that there shall never be more than 3 or 4 c. m. of water in the bottle. By regulating the water-cock, the proper supply of water may be determined.

The same apparatus may be used as an aspirator. When the instrument is to be used to draw air through any apparatus, the upper end of the tube *a b* is closed with a cork, and the tube *e f* is connected with the apparatus through which the current of air is to be drawn. The force of the current of air is to a certain degree affected by the size of the tube *a b*; to diminish the effective calibre of this tube, in case a gentle current of air is required, a glass rod as long as the tube may be passed down the tube through a cork inserted at *a*. The same apparatus may thus be made to produce a gentle or a powerful current of air.

In default of a blast-lamp, platinum crucibles may readily be ignited in a fire of coke or anthracite. To this end, place the tightly covered platinum crucible in a somewhat larger crucible of refractory clay or Hessian ware, and pack the space between the two crucibles tightly with calcined magnesia, so that the platinum may nowhere come in contact with the clay. Cover the coarse crucible, and place it, with its contents, in the coal fire, in such a manner that it may be gradually heated; finally, imbed the crucible in the glowing coals and urge the draught of the furnace for half an hour. The degree of heat to which the contents of the platinum crucible may be exposed in this way in an efficient fire, is really far greater than that of the blast-lamps above described; but the lamps are more convenient than the fire.

The effect of a simple Bunsen's burner may be greatly increased without the use of any blower, by surrounding its flame with a cylinder of fire-clay, 3 inches in diameter by 4 or 5 inches high, and having walls at least $\frac{3}{8}$ of an inch thick. The crucible, or other body to be heated, is hung in the middle of this chimney, and is thus exposed not only to the direct heat of the flame, but also to the radiant heat from the clay walls which surround it.

Fig. 13.

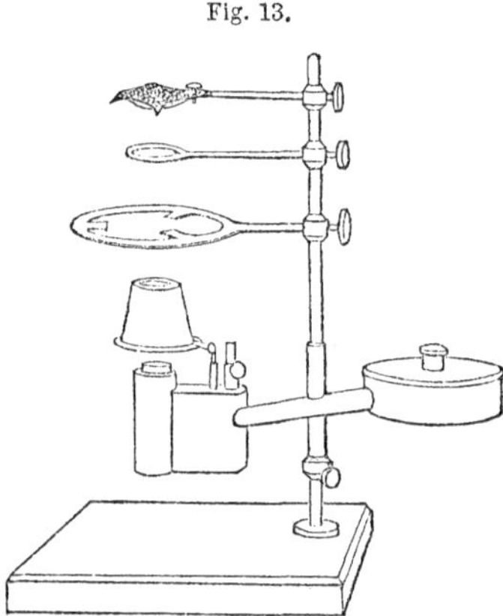

Where no gas is to be had, an alcohol lamp with circular wick, of some one of the numerous forms sold under the name of Berzelius's Argand Spirit Lamp (Fig. 13), will be found useful. These argand lamps are usually mounted on a lamp-stand provided with three brass rings; but the fittings of these lamps are all made slender, in order not to carry off too much heat. When it is necessary to heat heavy vessels, other supports must be used.

165. *Iron-stand, Tripod, Wire-gauze, and Triangle.* To support vessels over the gas-lamp, an iron-stand is used, consisting of a stout vertical rod fastened into a heavy, cast-iron foot, and three iron rings of graduated sizes secured to the vertical rod with binding screws ; all the rings may be slipped off the rod, or any ring may be adjusted at any convenient elevation. The general arrangement is not unlike that of the stand which makes part of the Berzelius lamp (Fig. 13), although simpler and cheaper. As a general rule, it is not best to apply the direct flame of the lamp to glass and porcelain vessels ; hence a piece of iron wire-gauze of medium fineness is stretched loosely over the largest ring, and bent downwards a little for the reception of round-bottomed vessels. On this gauze flasks and porcelain dishes are usually supported. Crucibles, or dishes, too small for the smallest ring belonging to the stand, are conveniently supported on an equilateral triangle made of three pieces of soft iron wire twisted together at the apices ; this triangle is laid on one of the rings of the stand. An iron tripod—that is, a stout ring supported on three legs—may often be used instead of the stand above described, but it is not so generally useful because of the difficulty of adjusting it at various heights ; with a sufficiency of wooden blocks wherewith to raise the lamp or the tripod as occasion may require, it may be made available.

166. *Water-bath and Sand-bath.* It is often necessary to evaporate solutions, or to dry precipitates at a moderate temperature which can permanently be kept below a certain known limit. Thus, when an aqueous solution is to be quietly evaporated without spirting or jumping, the temperature of the solution must never be suffered to rise above the boiling point of water, nor even quite to reach this point. This quiet evaporation is best effected by the use of a water-bath—a copper cup whose top is made of concentric rings of different diameters, to adapt it to dishes of various size (Fig. 14). This cup, two-thirds full of water, is supported on the iron-stand over the lamp, and the dish containing the solution to be evaporated is placed on that one of the several rings which will permit the greater part of the dish to sink into the copper cup. The steam rising from the water impinges upon the bottom of the dish, and brings the liquid within it to a temperature which insures the evaporation of the water, but will not cause any actual ebullition. The water in the copper cup must never be allowed to boil away. Wherever a constant supply of steam is at hand, as in buildings warmed by steam, the copper cup above described may be converted into a steam-bath, by attaching it to a steam-pipe by means of a small tube provided with a stop-cock.

Fig. 14.

A cheap but serviceable water-bath may be made from a quart milk-can, oil-can, tea-canister, or any similarly shaped tin vessel, by inserting the stem of a glass funnel into the neck of the can through a well-fitting cork. In this funnel the dish containing the liquor to be evaporated rests. The can contains the water, which is to be kept just boiling. On account of the shape of the funnel, dishes of various sizes can be used with the same apparatus.

When a gradual and equable heat, higher than can be obtained upon the water-bath, is required, a sand-bath will sometimes be found useful. A cheap and convenient sand-bath may be made by beating a disk of thin sheet-iron, about four inches in diameter, into the form of a saucer or shallow pan, and placing within it a small quantity of dry sand. The dish or flask to be heated is imbedded in the sand, and the apparatus placed upon a ring of the iron-stand over a gas-lamp.

167. *Blow-pipe.* The cheapest and best form of mouth blow-pipe for chemical purposes is a tube of tin-plate, about 18 c. m. long, 2 c. m. broad at one end, and tapering to 0.7 c. m. at the other (Fig. 15); the broad end is closed, and a little above this closed end a small cylindrical tube of brass about 5 c. m. long is soldered in at right angles; this brass tube is slightly conical at the end, and carries a small nozzle or tip, which may be made either of brass or platinum.

Fig. 15.

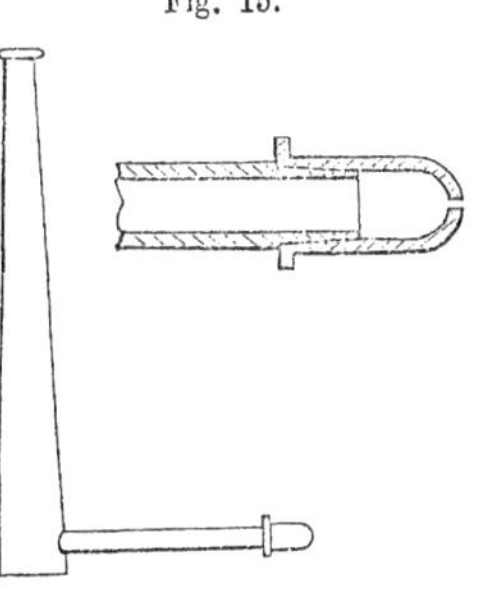

The tip should be drilled out of a solid piece of metal, and should not be fastened upon the brass-tube with a screw. A trumpet-shaped mouth-piece of horn or boxwood is a convenient, though by no means essential addition to this blow-pipe.

The blow-pipe may be used with a candle, with gas, or with any hand-lamp proper for burning oil, petroleum, or any of the so-called *burning fluids*, provided that the form of the lamp below the wick-holder is such as to permit the close approach of the object to be heated to the side of the wick. When a lamp is used, a wick about 1.2 c. m. long and 0.5 c. m. broad is more convenient than a round or narrow wick; a wick of this sort, though hardly so wide, is used in some of the open burning-fluid (naphtha) lamps now in common use. The wick-holder should be filed off on its longer dimension a little obliquely, and the wick cut parallel to the holder, in order that the blow-pipe flame may be directed downwards when necessary (Fig. 16). A gas flame suitable for the blow-pipe is readily obtained by slipping a narrow brass-tube, open at both ends, into the tube *f* of Bunsen's burner. (See Fig. 7.) This blow-pipe-tube must be long enough to close the air apertures in the tube *f*, and should be pinched together and filed off obliquely on top, as shown in Fig. 16 ; it may usually be obtained with the burner from dealers in chemical ware.

To use the mouth blow-pipe, place the open end of the tin tube between the lips, or, if the pipe is provided with a mouth-piece, press the trumpet-shaped mouth-piece against the lips ; fill the mouth with air till the cheeks are widely distended, and insert the tip in the flame of a lamp or candle ; close the communication between the lungs and the mouth, and force a current of air through the tube by squeezing the air in the mouth with the muscles of the cheeks, breathing, in the meantime, regularly and quietly through the nostrils. The knack of blowing a steady stream for several minutes at a time is readily acquired by a little practice. It will be at once observed that the appearance of the flame varies considerably, according to the strength of the blast and the position of the jet with reference to the wick.

Fig. 16.

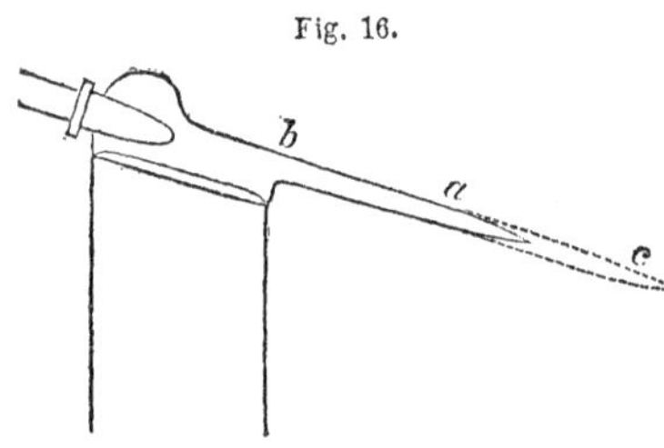

When the jet of the blow-pipe is inserted into the middle of a candle-flame, or is placed in the lamp-flame in the position shown in Fig. 16, and a strong blast is forced through the tube, a long, blue cone of flame, *a b*, is produced, beyond and outside of which stretches a more or less colored outer cone towards *c*. The point of greatest heat in this flame is at the point of the inner blue cone, because the combustible gases are there supplied with just the quantity of oxygen necessary to consume them, but between this point and the extremity of the flame the combustion is concentrated and intense. The greater part of the flame thus produced is *oxidizing* in its effect, and this flame is technically called the *oxidizing flame.* From the point *a* of the inner blue cone, the heat of the flame diminishes in both directions, towards *b* on the one hand, and towards *c* on the other ; most substances require the temperature to be found between *a* and *c*. Oxidation takes place most rapidly at, or just beyond, the point *c* of the flame, provided that the temperature at this point is high enough for the special substance to be heated.

A flame of precisely the opposite chemical effect may be produced with the blow-pipe. To obtain a good *reducing* flame, it is necessary to place the tip of the blow-pipe, not within, but just outside of the flame, and to blow rather over than through the middle of the flame (Fig. 17). In this manner the flame is less altered in its general character than in the former case, the chief part consisting of a large, luminous cone, containing a quantity of free carbon in a state of intense ignition, and just in the condition for taking up oxygen. This flame is, therefore, *re-*

Fig. 17.

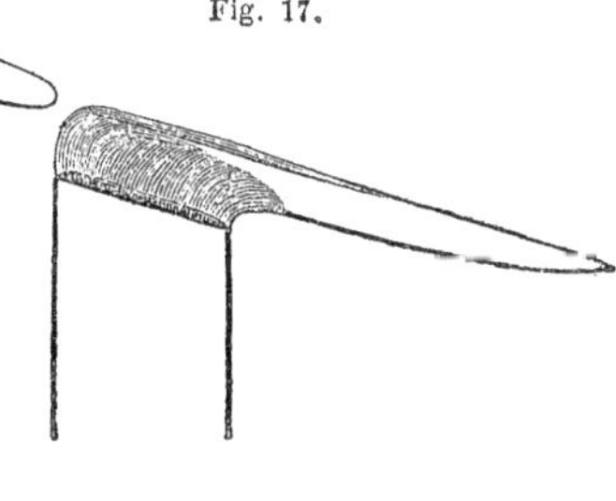

ducing in its effect, and is technically called the *reducing* flame. The substance which is to be reduced by exposure to this flame, should be completely covered up by the luminous cone, so that contact with the air may be entirely avoided. It is to be observed that, whereas to produce an effective oxidizing flame a strong blast of air is desirable, to get a good reducing flame the operator should blow gently, with only enough force to divert the lamp-flame.

Substances to be heated in the blow-pipe flame are supported, sometimes on charcoal, and sometimes on platinum foil or wire, or in platinum spoons or forceps. Charcoal is especially suitable for a support in experiments of reduction.

168. *Platinum foil and wire. Pincers.* A piece of platinum foil about 1½ inches long and 1 inch wide will be sufficient. The foil should be at least so thick that it does not crinkle when wiped; and it is more economical to get foil which is too thick than too thin, for it requires frequent cleaning. To keep foil in good order it should be frequently scoured with fine moist sand, and in case the foil becomes wrinkled it may be burnished by placing it upon the bottom of an inverted agate or porcelain mortar and rubbing it strongly with the pestle.

A bit of platinum wire, not stouter than the wire of a small pin and about 3 inches long, will last a long time with careful usage. It may be cleaned by long-continued boiling in water. A small loop about as large as this O, should be bent at each end of the wire.

When platinum foil is to be heated it may be held at one end with a pair of the small steel pincers known as jewellers' tweezers. A piece of platinum wire, as long as the one above described, can be held in the fingers without inconvenience, for platinum is, comparatively speaking, a bad conductor of heat. Pieces of wire too short to be held may be made serviceable by thrusting one end of the wire into the end of a glass rod or closed tube which has been softened in the blow-pipe flame.

169. *Platinum Crucibles.* For several of the operations of quantitative analysis as now practised, platinum crucibles are indispensable, and though not absolutely necessary for the profitable study of qualitative analysis, one of these vessels will often be found convenient by the student of the elements of analysis. It will be well, therefore, for the student who proposes to continue his chemical studies beyond qualitative analysis, to procure a platinum crucible once for all. A crucible of the capacity of about 20 cubic centimetres will be large enough for most uses ; it should be cylindrical rather than flaring, and should be provided with a loose cover in the form of a shallow dish.

No other metal, and no mixture of substances from which a metal can be reduced, must ever be heated in a platinum crucible, or upon platinum foil or wire, for platinum forms alloys with other metals which are much more fusible than itself. If once alloyed with a baser metal, the platinum ceases to be applicable to its peculiar uses.

Platinum may be cleaned by boiling it in either nitric or chlorhydric acid, by fusing acid sulphate of sodium upon it, or by scouring it with fine sand. Aqua-regia and chlorine-water dissolve platinum; the sulphides, cyanides, and hydrates of sodium and potassium, when fused in platinum vessels, slowly attack the metal.

170. *Wash-bottle.* A wash-bottle is a flask with a uniformly thin bottom, closed with a sound cork or caoutchouc stopper through which pass two glass tubes, as shown in Fig. 18.

Fig. 18.

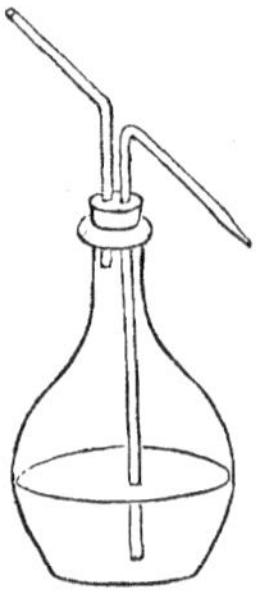

The outer end of the longer tube is drawn to a moderately fine point. A short bend near the bottom of this longer tube, in the same plane and direction as the upper bend, is of some use, because it enables the operator to empty the flask more completely by inclining it. By blowing into the short tube, a stream of water will be driven out of the long tube with considerable force. This force with which the stream is projected adapts the apparatus to removing precipitates from the sides of vessels as well as to washing them on filters. For use in analytical operations, it is often convenient to attach a caoutchouc tube 12 or 15 c. m. long to the tube through which the air is blown ; this flexible tube should be provided with a glass mouth-piece, consisting of a bit of glass tubing about 3 c. m. long. As the wash-bottle is often filled with hot or even boiling water, it may be improved by binding about its neck a ring of cork, or winding the neck closely with smooth cord. It may then be handled without inconvenience when hot.

The method of making a wash-bottle is described in the following paragraphs.

171. *Glass Tubing.* Two qualities of glass tubing are used in chemical experiments : that which softens readily in the flame of a gas or spirit-lamp, and that which fuses with extreme difficulty in the flame of the blast-lamp. These two qualities are distinguished by the terms *soft* and *hard* glass. Soft glass is to be preferred for all uses except the intense heating, or ignition, of dry substances.

Fig. 19 represents the common sizes of glass tubing, both hard and soft, and shows also the proper thickness of the glass walls for each size. The numbers ranging from 4 to 8 are best suited for use in qualitative analysis.

Fig. 19.

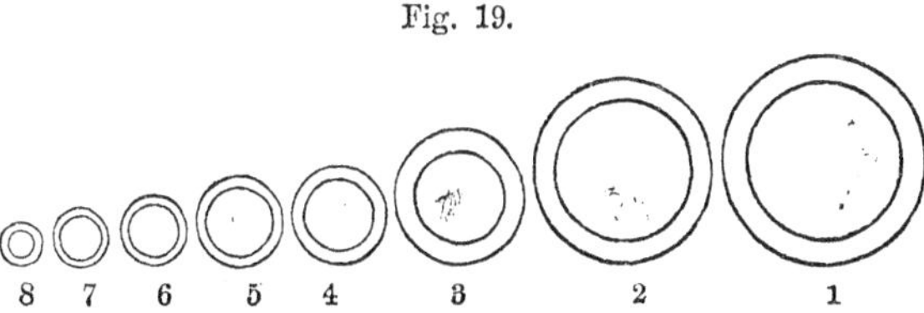

172. *Stirring Rods.* Cut a short stick of glass rod, No. 8 or 7, into pieces four or five inches long (see the next paragraph), and round the sharp ends by fusion in the blow-pipe-flame.

173. *Cutting and cracking glass.* Glass tubing and glass rod must generally be cut to the length required for any particular apparatus. A sharp triangular file is used for this purpose. The stick of tubing, or rod, to be cut is laid upon a table, and a deep scratch is made with the file at the place where the fracture is to be made. The stick is then grasped with the two hands, one on each side of the mark, while the thumbs are brought together just at the scratch. By pushing with the thumbs and pulling in the opposite direction with the fingers, the stick is broken squarely at the scratch, just as a stick of candy or dry twig may be broken. The sharp edges of the fracture should invariably be made smooth, either with a wet file, or by softening the end of the tube or rod in the lamp (§ 174). Tubes or rods of sizes four to eight inclusive may readily be cut in this manner; the larger sizes are divided with more difficulty, and it is often necessary to make the file-mark both long and deep. An even fracture is not always to be obtained with large tubes. The lower ends of glass funnels, and those ends of gas delivery-tubes which enter the bottle or flask in which the gas is generated, should be filed off or ground off on a grindstone, obliquely (Fig. 20), to facilitate the dropping of liquids from such extremities.

Fig. 20.

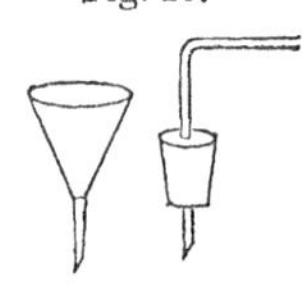

In order to cut glass plates, the glazier's diamond must be resorted to. For the cutting of exceedingly thin glass tubes, and of other glass ware, like flasks, retorts, and bottles, still other means are resorted to, based upon the sudden and unequal application of heat. The process divides itself into two parts,—the producing of a crack in the required place, and the subsequent guiding of this crack in the desired direction. To produce a crack, a scratch must be made with the file, and to this scratch a pointed bit of red-hot charcoal, or the jet of flame produced by the mouth blow-pipe, or a very fine gas-flame, or a red-hot glass-rod, or iron wire, may be applied. If the heat does not produce a crack, a wet stick or file may be touched upon the hot spot. Upon any part of a glass surface except the edge, it is not possible to control perfectly the direction and extent of this first crack ; at an edge a small crack may be started with tolerable certainty by carrying the file mark entirely *over* the edge. To guide the crack thus started, a pointed bit of charcoal or slow-match may be used. The hot point must be kept on the glass from 1 c. m. to 0.5 c. m. in advance of the point of the crack. The crack will follow the hot point, and may therefore be carried in any desired direction. By turning and blowing upon the coal or slow-match, the point may be kept sufficiently hot. Whenever the place of experiment is supplied with common illuminating gas, a very small jet of burning gas may be advantageously substituted for the hot coal or slow-match. To obtain such a sharp jet, a piece of hard glass tube, No. 5, 10 c. m. long, and drawn to a very fine point (§ 174), should be placed in the caoutchouc tube, which ordinarily delivers the gas to the gas-lamp, and the gas should be lighted at the fine extremity. The burning jet should have a fine point, and should not exceed 1.5 c. m. in length. By a judicious use of these simple tools, broken tubes, beakers, flasks, retorts, and bottles may often be made to yield very useful articles of apparatus. No sharp edges should be allowed to remain upon glass apparatus. The durability of the apparatus itself, and of the corks and caoutchouc stoppers and tubing used with it, will be much greater, if all sharp edges are removed with the file, or still better, rounded in the lamp.

174. *Bending and closing glass tubes.* Tubing of sizes four to eight inclusive, can generally be worked in the common gas or spirit-lamp; for larger tubes the blast-lamp is necessary (§ 164). Glass tubing must not be introduced suddenly into the hottest part of the flame, lest it crack. Neither should a hot tube be taken from the flame and laid at once upon a cold surface. Gradual heating and gradual cooling are alike necessary, and are the more essential the thicker the glass; very thin glass will sometimes bear the most sudden changes of temperature, but thick glass and glass of uneven thickness absolutely require slow heating and annealing. When the end of a tube is to be heated, as in rounding sharp edges, more care is required in consequence of the great facility with which cracks start at an edge. A tube should, therefore, always be brought first into the current of hot air beyond the actual flame of the gas or spirit-lamp, and there thoroughly warmed, before it is introduced into the flame itself. If a blast-lamp is employed, the tube may be warmed in the smoky flame, before the blast is turned on, and may subsequently be annealed in the same manner ; the deposited soot will be burnt off in the first instance, and in the last, may be wiped off when the tube is cold. In heating a tube, whether for bending, drawing, or closing, the tube must be *constantly* turned between the fingers, and also moved a little to the right and left, in order that it may be uniformly heated all round, and that the temperature of the neighboring parts may be duly raised. If a tube, or rod, is to be heated at any part but an end, it should be held between the thumb and first two fingers of each hand in such a manner that the hands shall be below the tube or rod, with the palms upward, while the lamp-flame is between the hands. When the end of a tube or rod is to be heated, it is best to begin by warming the tube or rod about 2 c. m. from the end, and from thence to proceed slowly to the end.

The best glass will not be blackened or discolored during heating. Blackening occurs in glass which, like ordinary flint glass, contains oxide of lead as an ingredient. Glass containing much of this oxide is not well adapted for chemical uses. The blackening may sometimes be removed by putting the glass in the upper or outer part of the flame, where the reducing gases are consumed, and the air has the best access to the glass. The blackening may be altogether avoided by always keeping the glass in the oxidizing part of the flame.

Glass begins to soften and bend below a visible red

heat. The condition of the glass is judged of as much by the fingers as the eye ; the hands feel the yielding of the glass, either to bending, pushing, or pulling, better than the eye can see the change of color or form. It may be bent as soon as it is hot enough to yield in the hands, but can be drawn out only when much hotter than this. Glass tubing, however, should not be bent at too low a temperature ; the curves made at too low a heat are apt to be flattened, of unequal thickness on the convex and concave sides, and brittle.

In bending tubing to make gas-delivery tubes and the like, attention should be paid to the following points : 1st, the glass should be equally hot on all sides ; 2d, it should not be twisted, pulled out, or pushed together during the heating ; 3d, the bore of the tube at the bend should be kept round, and not altered in size ; 4th, if two or more bends be made in the same piece of tubing (Fig. 21, *a*), they should all be in the same plane, so that the finished tube will lie flat upon the level table.

Fig. 31.

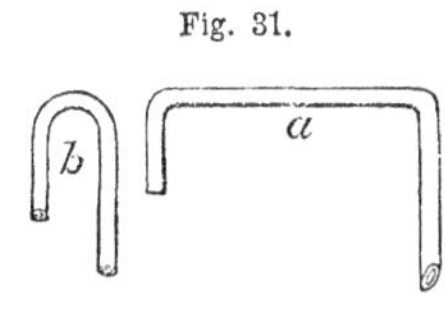

When a tube or rod is to be bent or drawn close to its extremity, a temporary handle may be attached to it by softening the end of the tube or rod, and pressing against the soft glass a fragment of glass tube, which will adhere strongly to the softened end. The handle may subsequently be removed by a slight blow, or by the aid of a file. If a considerable bend is to be made, so that the angle between the arms will be very small or nothing, as in a siphon, for example, the curvature cannot be well produced at one place in the tube, but should be made by heating, progressively, several centimetres of the tube, and bending continuously from one end of the heated portion to the other

(Fig. 21, *b*). Small and thick tube may be bent more sharply than large or thin tube.

In order to draw a glass tube down to a finer bore, it is simply necessary to thoroughly soften on all sides one or two centimetres' length of the tube, and then taking the glass from the flame, pull the parts asunder by a cautious movement of the hands. The larger the heated portion of glass, the longer will be the tube thus formed. Its length and fineness also increase with the rapidity of motion of the hands. If it is desirable that the finer tube should have thicker walls in proportion to its bore than the original tube, it is only necessary to keep the heated portion soft for two or three minutes before drawing out the tube, pressing the parts slightly together the while. By this process the glass will be thickened at the hot ring.

To obtain a tube closed at one end, it is best to take a piece of tubing, open at both ends, and long enough to make two closed tubes. In the middle of the tube a ring of glass, narrow as possible, must be made thoroughly soft. The hands are then separated a little, to cause a contraction in diameter at the hot and soft part. The point of the flame must now be directed, not upon the narrowest part of the tube, but upon what is to be the bottom of the closed tube. This point is indicated by the line *a* in Fig. 22. By withdrawing the right hand, the narrow part of the tube is attenuated, and finally melted off, leaving both halves of the original tube closed at one end, but not of the same form ; the right-hand half is drawn out into a long point, the other is more roundly closed. It is not possible to close handsomely the two pieces at once. The tube is seldom per-

Fig. 22.

fectly finished by the operation ; a superfluous knob of glass generally remains upon the end. If small, it may be got rid of by heating the whole end of the tube, and blowing moderately with the mouth into the open end. The knob, being hotter, and therefore softer than any other part, yields to the pressure from within, spreads out and disappears. If the knob is large, it may be cut off with scissors while red hot, or drawn off by sticking to it a fragment of tube, and then softening the glass above the junction. The same process may be applied to the too pointed end of the right-hand half of the original tube, or to any misshapen result of an unsuccessful attempt to close a tube, or to any bit of tube which is too short to make two closed tubes. When the closed end of a tube is too thin, it may be strengthened by keeping the whole end at a red heat for two or three minutes, turning the tube constantly between the fingers. It may be said in general of all the preceding operations before the lamp, that success depends on keeping the tube to be heated in constant rotation, in order to secure a uniform temperature on all sides of the tube.

175. *Blowing Bulbs.* Bulb-tubes, like the one represented in Fig. 23, are employed for reducing substances capable of forming sublimates upon the cold walls of the tube. They are readily made from bits of tubing, in the flame of Bunsen's burner, or in the common blow-pipe flame.

Fig. 23.

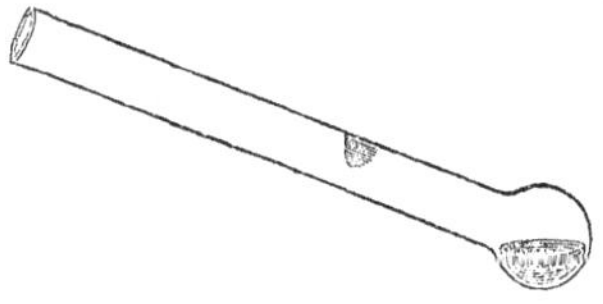

If the bulb desired is large in proportion to the size of the tube on which it is to be made, the walls of the tube must be thickened by rotation in the flame before the bulb can be blown. The thickened portion of the glass is then to be heated to a cherry-red, suddenly withdrawn from the flame, and expanded while hot by steadily blowing, or rather pressing, air into the tube with the mouth ; the tube must be constantly turned on its axis, not only while in the flame, but also while the bulb is being blown. If too strong or too sudden a pressure be exerted with the mouth, the bulb will be extremely thin, and quite useless. By watching the expanding glass, the proper moment for arresting the pressure may usually be determined. If the bulb obtained be not large enough, it may be reheated and enlarged by blowing into it again, provided that a sufficient thickness of glass remain.

If a bulb is to be blown in the middle of a piece of tubing, the thickening is effected by gently pressing the ends of the tube together while the glass is red-hot in the place where the bulb is to be ; if a large bulb is to be placed at the end of a tube, this end is first closed, and then suitably thickened by pressing the hot glass up with a piece of metal until enough has been accumulated at the end.

It is sometimes necessary to make a hole in the side of a tube or other thin glass apparatus. This may be done by directing a pointed flame from the blast-lamp upon the place where the hole is to be, until a small spot is red-hot, and then blowing forcibly into one end of the tube while the other end is closed by the finger ; at the hot spot the glass is blown out into a thin bubble, which bursts or may be easily broken off, leaving an aperture in the side of the tube.

It is hoped that these few directions will enable the attentive student to perform, sufficiently well, all the manipulations with glass tubes which the experiments described in this manual require. Much practice will alone give a perfect mastery of the details of glass-blowing.

176. *Caoutchouc.* Vulcanized caoutchouc is a most useful substance in the laboratory, on account of its elasticity, and because it resists so well most of the corrosive substances with which the chemist deals. It is used in three forms: (1) in tubing of various diameters comparable with the sizes of glass tubing; (2) in stoppers of various sizes, to replace corks; (3) in sheets. Caoutchouc tubing may be used to conduct all gases and liquids which do not corrode its substance, provided that the pressure under which the gas or liquid flows be not greater, or their temperature higher, than the texture of the tubing can endure. The flexibility of the tubing is a very obvious advantage in a great variety of cases. Short pieces of such tubing, a few centimetres in length, are much used, under the name of connectors, to make flexible joints in apparatus of which glass tubing forms part; flexible joints add greatly to the durability of such apparatus, because long glass tubes bent at several angles and connected with heavy objects, like globes, bottles or flasks full of liquid, are almost certain to break, even with the most careful usage; gas-delivery tubes, and all considerable lengths of glass tubing, should invariably be divided at one or more places, and the pieces joined again with caoutchouc connectors. The ends of glass tubing to be thus connected should be squarely cut, and then rounded in the lamp, in order that no sharp edges may cut the caoutchouc; the internal diameter of the caoutchouc tube must be a little smaller than the external diameter of the glass tubes; the slipping on of the connector is facilitated by wetting the glass. In some cases of delicate quantitative manipulations, in which the tightest possible joints must be secured, the caoutchouc connector is bound on to the glass tube with a silk or smooth linen string; the string is passed as tightly as possible twice round the connector, and tied with a square knot; the string should be moistened in order to prevent it from slipping while the knot is tying.

Caoutchouc stoppers of good quality are much more durable than corks, and are in every respect to be preferred. The German stoppers are of excellent shape and quality; the American, being chiefly intended for wine bottles, are apt to be too conical. Caoutchouc stoppers can be bored like corks (see the next section) by means of suitable cutters, and glass tubes can be fitted into the holes thus made with a tightness unattainable with corks. German stoppers may be bought already provided with one, two, and three holes. It is not well to lay in a large stock of caoutchouc stoppers; for, though they last a long time when in constant use, they not infrequently deteriorate when kept in store, becoming hard and somewhat brittle with age. These stoppers must not be confounded with the very inferior caps which were in use a few years ago.

Pieces of thin sheet caoutchouc are very conveniently used for making tight joints between large tubes of two different sizes, or between the neck of a flask, or bottle, and a large tube which enters it, or between the neck of a retort and the receiver into which it enters. A sufficiently broad and long piece of sheet caoutchouc is considerably stretched, wrapped tightly over the glass parts adjoining the aperture to be closed, and secured in place by a string wound closely about it and tied with a square knot.

177. *Corks.* It is often very difficult to obtain sound, elastic corks, of fine grain and of size suitable for large flasks and wide-mouthed bottles. On this account, bottles with mouths not too large to be closed with a cork cut across the grain should be chosen for chemical uses in preference to bottles which require large corks or bungs cut with the grain, and therefore offering continuous channels for the passage of gases, or even liquids. The kinds sold as champagne corks and as satin corks for phials, are suitable for chemical use. The best corks generally need to be softened before using; this softening may be effected by rolling the cork under a board upon the table, or under the foot upon the clean floor, or by gently squeezing it on all sides with the well-known tool expressly adapted for this purpose, and thence called a cork-squeezer ; steaming also softens the hardest corks.

Corks must often be cut with cleanness and precision ; a sharp, thin knife, such as shoemakers use, is desirable for this purpose. When a cork has been pared down to reduce its diameter, a flat file may be employed in finishing ; the file must be fine enough to leave a smooth surface upon the cork ; in filing a cork, a cylindrical, rather than a conical, form should be aimed at.

In boring holes through corks to receive glass tubes, a hollow cylinder of sheet brass sharpened at one end is a very convenient tool. Fig. 24 represents a set of such little cylinders of graduated sizes, slipping one within the other into a very compact form ; a stout wire, of the same length as the cylinders, accompanies the set, and serves a double purpose—passed transversely through two holes in the cap which terminates each cylinder, it gives the hand a better grasp of the tool while penetrating the cork ; and when the hole is made, the wire thrust through an opening in the top of the cap expels the little cylinder of cork which else would remain in the cutting cylinder of brass. That cutter, whose diameter is next below that of the glass tube to be inserted in the cork, is always to be selected, and if the hole it makes is too small, a round file must be used to enlarge the aperture; the round file, also, often comes in play to smooth the rough sides of a hole made by a dull cork-borer. A pair of small calipers is a very convenient, though by no means essential, tool in determining which size of cutter to employ. A flask which presents sharp or rough edges at the mouth can seldom be tightly corked, for the cork cannot be introduced into the neck without being cut or roughened; such sharp edges must be rounded in the lamp. In thrusting glass tubes through bored corks, the following directions are to be observed: (1.) The end of the tube must not present a sharp edge capable of cutting the cork. (2.) The tube should be grasped very close to the cork, in order to escape cutting the hand which holds the cork, should the tube break; by observing this precaution, the chief cause of breakage, viz., irregular lateral pressure, will be at the same time avoided. (3.) A funnel-tube must never be held by the funnel in driving it through a cork, nor a bent tube grasped at the bend, unless the bend comes immediately above the cork. (4.) If the tube goes very hard through the cork, the application of a little soap and water will facilitate its passage; but if soap is used, the tube can seldom be withdrawn from the cork after the latter has become dry. (5.) The tube must not be pushed straight into the cork, but

Fig. 24.

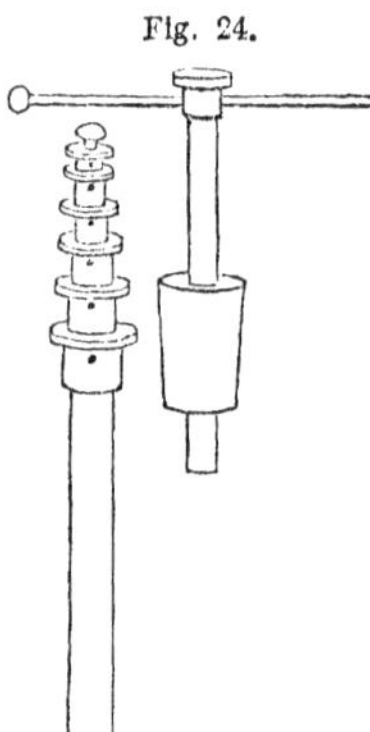

screwed in, as it were, with a slow rotary as well as onward motion. Joints made with corks should always be tested before the apparatus is used, by blowing into the apparatus, and at the same time stopping up all legitimate outlets. Any leakage is revealed by the disappearance of the pressure created. To the same end, air may be sucked out of an apparatus and its tightness proved by the permanence of the partial vacuum. To attempt to use a leaky cork is generally to waste time and labor, and to insure the failure of the experiment.

178. *Gas-bottle.* Fig. 25 represents a gas-bottle fitted for evolving sulphuretted hydrogen, carbonic acid, and other gases which can be prepared without heat. A straight glass tube of convenient length is slipped into the caoutchouc connector at the right to carry the gas into the solution to be tested. The neck of the bottle should be rather narrow, since it is difficult to obtain tight stoppers for bottles with wide mouths, but must nevertheless be wide enough to admit a cork, or better a caoutchouc stopper, capable of carrying both the delivery and the thistle tubes.

Fig. 25.

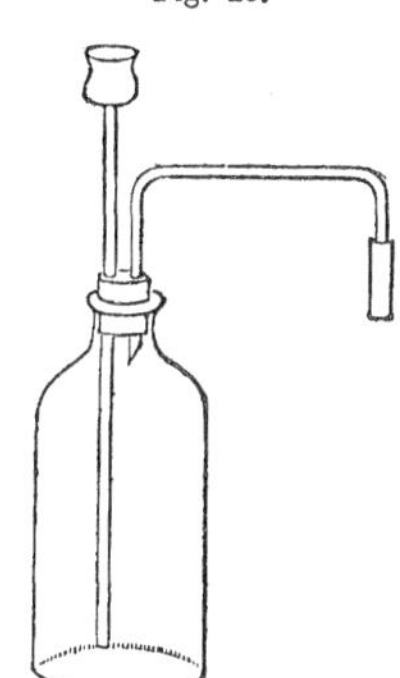

To prepare, for example, sulphuretted hydrogen gas, put a tablespoonful of fragments of sulphide of iron in the bottom of the bottle, replace the cork with its tubes, and press, or rather twist, it tightly into the neck of the bottle ; pour in enough water through the thistle-tube to seal the lower end of that tube, and finally as much concentrated sulphuric acid as would be equal to a tenth or a twelfth of the volume of the water.

At the start it is well thus to mix strong acid with the water in the bottle, for the heat generated by the union of the two liquids serves to warm the apparatus, and to facilitate the decomposition of the sulphide of iron ; but it must be remembered that strong sulphuric acid is by itself unfit for generating sulphuretted hydrogen, and that the evolution of gas would be checked if much of it were added. When the flow of gas ceases, pour a small portion of dilute sulphuric acid into the thistle-tube, and repeat this operation as often as may be necessary to maintain a constant stream of gas. Dilute acid fit for this purpose may be prepared by mixing 1 volume of strong sulphuric acid with 14 volumes of water; the water should be well stirred and the acid poured into it in a fine stream.

In precipitating the members of Classes II and III with sulphuretted hydrogen, the gas delivery-tube should not dip deeper than about an inch beneath the surface of the liquid in the beaker. A rapid current of gas is useless and wasteful. The best method of operating is to pour dilute sulphuric acid into the thistle-tube in such quantity that the bubbles of gas may follow one another slowly enough to be counted without effort.

179. *Self-regulating Gas-generator.* An apparatus which is always ready to deliver a constant stream of sulphuretted hydrogen, and yet does not generate the gas except when it is immediately wanted for use, is a great convenience in an active laboratory. The same remark applies to the two gases, hydrogen and carbonic acid, which are likewise used in considerable quantities in quantitative analysis, and which can be conveniently generated in precisely the same form of apparatus which is advantageous for sulphuretted hydrogen. Such a generator may be made of divers dimensions. The following directions, with the accompanying figure (Fig. 26), will enable the student to construct an apparatus of convenient size. Procure a glass cylinder, 20 or 25 c. m. in diameter, and 30 or 35 c. m. high ; ribbed candy jars are sometimes to be had of about this size ; procure also a stout tubulated bell-glass, 10 or 12 c. m. wide, and 5 or 7 c. m. shorter than the cylinder. Get a basket of sheet-lead, 7.5 c. m. deep, and 2.5 c. m. narrower than the bell-glass, and bore a number of small holes in the sides and bottom of this basket. Cast a circular plate of lead, 7 m. m. thick, and of a diameter 4 c. m. larger than that of the glass cylinder ; on what is intended for its under side, solder three equidistant leaden strips, or a continuous ring of lead, to keep the plate in proper position as a cover for the cylinder. Fit tightly to each end of a good brass gas-cock a piece of brass tube 8 c. m. long, 1.5 to 2 c. m. wide, and stout in metal. Perforate the centre of the leaden plate, so that one of these tubes will snugly pass through the orifice, and secure it by solder, leaving 5 c. m. of the tube projecting below the plate. Attach to the lower end of this tube a stout hook on which to hang the leaden basket. By means of a sound cork and common sealing-wax, or a cement made of oil mixed with red and white lead, fasten this tube into the tubulature of the bell-glass air-tight, and so firmly that the joint will bear a weight of several pounds. Hang the basket by means of copper wire

Fig. 26.

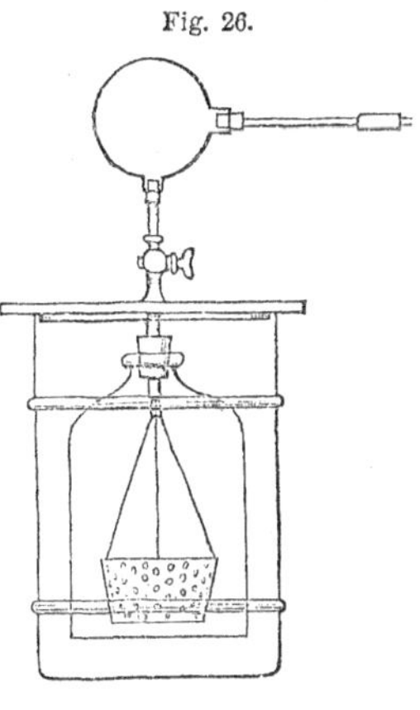

within the bell, 5 c. m. above the bottom of the latter. To the tube which extends above the stop-cock attach by a good cork the neck of a tubulated receiver, of 100 or 150 c. c. capacity, the interior of which has been loosely stuffed with cotton. Into the second tubulature of the receiver fit tightly the delivery-tube carrying a caoutchouc connector ; into this connector can be fitted a tube adapted to convey the gas in any desired direction. When many persons use the same generator, each person must bring his own tube.

To charge the apparatus, fill the cylinder with dilute acid to within 10 or 12 c. m. of the top ; fill the basket with fragments of sulphide of iron ; hang the basket in the bell, and put the bell-glass full of air into its place, with the stop-cock closed. On opening the cock, the weight of the acid expels the air from the bell, the acid comes in contact with the solid in the basket, and a steady supply of gas is generated until either the acid is saturated or the solid dissolved ; if the cock be closed, the gas accumulates in the bell, and pushes the acid below the basket, so that all action ceases. In cold weather, the apparatus must be kept in a warm place. For generating sulphuretted hydrogen, sulphuric acid, diluted with fourteen parts of water, is used ; for hydrogen, zinc and sulphuric acid, diluted with four or five parts of water ; while for carbonic acid, chalk and muriatic acid, diluted with two or three parts of water, should be taken.

180. *Mortars.* Whenever the substance to be analyzed occurs in the form of large pieces or coarse powder, it should, as a general rule, be pulverized by mechanical means before subjecting it to the action of solvents. Mortars of iron, steel, agate, or porcelain are used for this purpose, according to the character of the substance to be powdered.

An iron mortar is useful for coarse work, and for effecting the first rough breaking up of substances which are subsequently powdered in the agate or porcelain mortar. If there be any risk of fragments being thrown out of the mortar, it should be covered with a cloth or piece of stiff paper, having a hole in the middle through which the pestle may be passed. Instead of the common iron mortar, a small steel mortar, of the kind called diamond mortars by dealers in chemical ware, may be used for crushing minerals. Pieces of stone, minerals, and lumps of brittle metals may be safely broken into fragments suitable for the mortar by wrapping them in strong paper, laying them so enclosed upon an anvil, and striking them with a heavy hammer. The paper envelope retains the broken particles, which might otherwise fly about in a dangerous manner, and be lost.

The best porcelain mortars are those known by the name of Wedgewood-ware, but there are many cheaper substitutes. Porcelain mortars will not bear sharp and heavy blows; they are intended rather for grinding or triturating saline substances than for hammering; the pestle may either be formed of one piece of porcelain, or a piece of porcelain cemented to a wooden handle; the latter is the less desirable form of pestle. Unglazed porcelain mortars are to be preferred. In selecting mortars, the following points should be attended to: 1st, the mortar should not be porous; it ought not to absorb strong acids or any colored fluid, even if such liquids be allowed to stand for hours in the mortar; 2d, it should be very hard, and its pestle should be of the same hardness; 3d, it should be sound; 4th, it should have a lip, for the convenience of pouring out liquids and fine powders. As a rule, porcelain mortars will not endure sudden changes of temperature. They may be cleaned by rubbing in them a little sand soaked in nitric or sulphuric acid, or, if acids are not appropriate, in caustic soda.

Agate mortars are only intended for trituration ; a blow would break them. They are exceedingly hard and impermeable. The material is so precious and so hard to work, that agate mortars are always small. The pestles are generally inconveniently short, a difficulty which may be remedied by fitting the agate pestle into a wooden handle.

In all grinding operations in mortars, whether of porcelain or agate, it is expedient to put only a small quantity of the substance to be powdered into the mortar at once. The operation of powdering will be facilitated by sifting the matter as fast as it is powdered, returning to the mortar the particles which are too large to pass through the sieve.

181. *Spatulæ.* For transferring substances in powder, or in small grains or crystals, from one vessel to another, spatulæ and scoops, made of horn or bone, are convenient tools. A coarse bone paper-knife makes a good spatula for laboratory use. Cards, free from glaze and enamel, are excellent substitutes for spatulæ.

INDEX.

ABBOT (H. L.) Siege Artillery in the Campaign against Richmond, with Notes on the 15-inch Gun, including an Algebraic Analysis of the Trajectory of a Shot in its ricochet upon smooth water. Illustrated with detailed drawings of the U. S. and Confederate rifled projectiles. By Henry L. Abbot, Maj. of Engineers, and Bvt. Maj.-General U. S. Vols., commanding Siege Artillery, Armies before Richmond. Paper No. 14, Professional Papers, Corps of Engineers. 1 vol. 8vo, cloth....... $3.50

ALEXANDER (J. H.) Universal Dictionary of Weights and Measures, Ancient and Modern, reduced to the standards of the United States of America. By J. H. Alexander. New edition, enlarged. 1 vol. 8vo, cloth. 3.50

BENET (S. V.) Electro-Ballistic Machines, and the Schultz Chronoscope. By Brevet Lieut.-Colonel S. V. Benét. 1 vol. 4to, illustrated, cloth...................... 4.00

BROOKLYN WATER WORKS. Containing a Descriptive Account of the Construction of the Works, and also Reports on the Brooklyn, Hartford, Belleville, and Cambridge Pumping Engines With illustrations. 1 vol. folio, cloth.............................. 15.00

BURGH (N. P.) Modern Marine Engineering, applied to Paddle and Screw Propulsion. Consisting of 36 colored plates, 259 Practical Woodcut Illustrations, and 403 pages of Descriptive Matter, the whole being an exposition of the present practice of the following firms: Messrs. J. Penn & Sons; Messrs. Maudslay, Sons & Field; Messrs. James Watt & Co.; Messrs. J. & G. Rennie; Messrs. R. Napier & Sons; Messrs. J. & W. Dudgeon; Messrs. Ravenhill & Hodgson; Messrs. Humphreys & Tenant; Mr. J. T. Spencer, and Messrs. Forrester & Co. By N. P. Burgh, Engineer. In 1 thick vol. 4to, cloth............................. 30.00

Do. do. half morocco..................... 35.00

CAMPIN on the Construction of Iron Roofs. By Francis Campin. 8vo, with plates, cloth.................. $3.00

CHAUVENET (Prof. Wm.) New Method of Correcting Lunar Distances, and Improved Method of Finding the Error and Rate of a Chronometer, by equal altitudes. By Wm. Chauvenet, LL. D. 1 vol. 8vo, cloth..... 2.00

CLOUGH (A. B.) Contractors' Manual and Builders' Price-Book. By A. B. Clough, Architect. 1 vol. 18mo, cloth.................................... 75

COLBURN (Zerah). Engineering—an Illustrated Weekly Journal, conducted by Zerah Colburn, London. 32 pages folio. Is promptly received here by weekly steamers. Price, per annum..................... 10.00

"This is the ablest Engineering paper published, and is edited by one of the best known scientific men of the day. It is finely and profusely illustrated, and printed in the best manner."

COLBURN. The Gas Works of London. By Zerah Colburn, C. E. 1 vol. 12mo, boards................. 75

CRAIG (B. F.) Weights and Measures. An Account of the Decimal System, with Tables of Conversion for Commercial and Scientific Uses. By B. F. Craig, M. D. 1 vol. square 32mo, limp cloth................ 50

"The most lucid, accurate, and useful of all the hand-books on this subject that we have yet seen. It gives forty-seven tables of comparison between the English and French denominations of length, area, capacity, weight, and the centigrade and Fahrenheit thermometers, with clear instructions how to use them; and to this practical portion, which helps to make the transition as easy as possible, is prefixed a scientific explanation of the errors in the metric system, and how they may be corrected in the laboratory."—*Nation.*

FRANCIS. Lowell Hydraulic Experiments, being a selection from Experiments on Hydraulic Motors, on the Flow of Water over Wiers, in Open Canals of Uniform Rectangular Section, and through submerged Orifices and diverging Tubes. Made at Lowell, Massachusetts. By James B. Francis, C. E. 2d edition, revised and enlarged, with many new experiments, and illustrated with twenty-three copperplate engravings. 1 vol. 4to, cloth...................................... 15.00

FRANCIS. On the Strength of Cast-Iron Pillars, with Tables for the use of Engineers, Architects, and Builders. By James B. Francis, Civil Engineer. 1 vol. 8vo, cloth.. $2.00

"A scientific treatise of inestimable value to those for whom it is intended."—*Boston Daily Advertiser.*

GILLMORE (Gen. Q. A.) Treatise on Limes, Hydraulic Cements, and Mortars. Papers on Practical Engineering, U. S. Engineer Department, No. 9, containing Reports of numerous Experiments conducted in New York City, during the years 1858 to 1861, inclusive. By Q. A. Gillmore, Bvt. Maj.-Gen, U. S. A., Major, Corps U. S. Engineers. With numerous illustrations. 1 vol. 8vo, cloth.............................. 4.00

"This work contains a record of certain experiments and researches made under the authority of the Engineer Bureau of the War Department from 1858 to 1861, upon the various hydraulic cements of the United States, and the materials for their manufacture The experiments were carefully made, and are well reported and compiled."—*Journal Franklin Institute.*

HARRISON. The Mechanics' Tool Book, with Practical Rules and Suggestions for Use of Machinists, Iron Workers, and others. By W. B. Harrison, associate editor of the "American Artisan." Illustrated with 44 engravings. 12mo, cloth........................ 2.50

HENRICI (Olaus). Skeleton Structures, especially in their application to the Building of Steel and Iron Bridges. By Olaus Henrici. With folding plates and diagrams. 1 vol. 8vo, cloth.............................. 3.00

HEWSON (Wm.) Principles and Practice of Embanking Lands from River Floods, as applied to the Levees of the Mississippi. By William Hewson, Civil Engineer. 1 vol. 8vo, cloth.............................. 2.00

"This is a valuable treatise on the principles and practice of embanking lands from river floods, as applied to Levees of the Mississippi, by a highly intelligent and experienced engineer. The author says it is a first attempt to reduce to order and to rule the design, execution, and measurement of the Levees of the Mississippi. It is a most useful and needed contribution to scientific literature."—*Philadelphia Evening Journal.*

HOLLEY (A. L.) Railway Practice. American and European Railway Practice, in the economical Generation of Steam, including the Materials and Construction of Coal-burning Boilers, Combustion, the Variable Blast, Vaporization, Circulation, Superheating, Supplying and Heating Feed-water, &c., and the Adaptation of Wood and Coke-burning Engines to Coal-burning; and in Permanent Way, including Road-bed, Sleepers, Rails, Joint-fastenings, Street Railways, &c., &c. By Alexander L. Holley, B. P. With 77 lithographed plates. 1 vol. folio, cloth $12.00

* * * "All these subjects are treated by the author in both an intelligent and intelligible manner. The facts and ideas are well arranged, and presented in a clear and simple style, accompanied by beautiful engravings, and we presume the work will be regarded as indispensable by all who are interested in a knowledge of the construction of railroads and rolling stock, or the working of locomotives."—*Scientific American.*

HUNT (R. M.) Designs for the Gateways of the Southern Entrances to the Central Park. By Richard M. Hunt. With a description of the designs. 1 vol. 4to, illustrated, cloth 5.00

KING (W. H.) Lessons and Practical Notes on Steam, the Steam Engine, Propellers, &c., &c., for Young Marine Engineers, Students, and others. By the late W. H. King, U. S. Navy. Revised by Chief Engineer J. W. King, U. S. Navy. Twelfth edition, enlarged. 8vo, cloth 2.00

"This is the twelfth edition of a valuable work of the late W. H. King, U. S. Navy. It contains lessons and practical notes on Steam and the Steam Engine, Propellers, &c. It is calculated to be of great use to young marine engineers, students, and others. The text is illustrated and explained by numerous diagrams and representations of machinery. This new edition has been revised and enlarged by Chief Engineer J. W. King, U. S. Navy, brother to the deceased author of the work."—*Boston Daily Advertiser.*

McCORMICK (R. C.). Arizona: Its Resources and Prospects. By Hon. R. C. McCormick. With map, 8vo, paper 25

MINIFIE (Wm.) Mechanical Drawing. A Text-Book of Geometrical Drawing for the use of Mechanics and Schools, in which the Definitions and Rules of Geometry are familiarly explained; the Practical Problems are arranged, from the most simple to the more complex, and in their description technicalities are avoided as much as possible. With illustrations for Drawing Plans, Sections, and Elevations of Buildings and Machinery; an Introduction to Isometrical Drawing, and an Essay on Linear Perspective and Shadows. Illustrated with over 200 diagrams engraved on steel. By Wm. Minifie, Architect. Seventh Edition. With an Appendix on the Theory and Application of Colors. 1 vol. 8vo, cloth $4.00

"It is the best work on Drawing that we have ever seen, and is especially a text-book of Geometrical Drawing for the use of Mechanics and Schools. No young Mechanic, such as a Machinist, Engineer, Cabinet-Maker, Millwright or Carpenter, should be without it."—*Scientific American.*

"One of the most comprehensive works of the kind ever published, and cannot but possess great value to builders. The style is at once elegant and substantial."—*Pennsylvania Inquirer.*

"We think this the best work on this subject, which is saying very much; as much attention has been given to the science of Drawing. There is nothing in the range of drawing that cannot be found in this book, and which is not well explained."—*Ohio Teacher.*

"Whatever is said is rendered perfectly intelligible by remarkably well-executed diagrams on steel, leaving nothing for mere vague supposition; and the addition of an introduction to isometrical drawing, linear perspective, and the projection of shadows, winding up with a useful index to technical terms."—*Glasgow Mechanics' Journal.*

☞ The British Government has authorized the use of this book in their schools of art at Somerset House, London, and throughout the Kingdom.

MINIFIE (Wm.) Geometrical Drawing. Abridged from the Octavo edition, for the use of Schools. Illustrated with 48 steel plates. Fifth edition, 1 vol. 12mo, half roan 1.50

"It is well adapted as a text-book of drawing to be used in our High Schools and Academies where this useful branch of the fine arts has been hitherto too much neglected."—*Boston Journal.*

PIERCE (Prof. Benj.) System of Analytical Mechanics. Physical and Celestial Mechanics, by Benjamin Pierce, Perkins Professor of Astronomy and Mathematics in Harvard University, and Consulting Astronomer of the American Ephemeris and Nautical Almanac. Developed in four systems of Analytical Mechanics, Celestial Mechanics, Potential Physics, and Analytic Morphology. 1 vol. 4to, cloth $10.00

"I have re-examined the memoirs of the great geometers, and have striven to consolidate their latest researches and their most exalted forms of thought into a consistent and uniform treatise. If I have hereby succeeded in opening to the students of my country a readier access to these choice jewels of intellect; if their brilliancy is not impaired in this attempt to reset them; if, in their own constellation, they illustrate each other, and concentrate a stronger light upon the names of their discoverers; and, still more, if any gem which I may have presumed to add is not wholly lustreless in the collection,—I shall feel that my work has not been in vain."—*Extract from the Preface.*

PLYMPTON. The Blow-Pipe; a System of Instruction in its Practical Use, being a Graduated Course of Analysis for the Use of Students and all those engaged in the Examination of Metallic Combinations. Second edition, with an appendix and a copious index. By Geo. W. Plympton, of the Polytechnic Institute, Brooklyn. 1 vol. 12mo, cloth 2.00

POOK (S. M.) Method of Comparing the Lines and Draughting Vessels Propelled by Sail or Steam. Including a chapter on Laying off on the Mould-Loft Floor. By Samuel M. Pook, Naval Constructor. 1 vol. 8vo, with illustrations, cloth 5.00

"Few men have had better opportunity to study the marine architecture of the past and present than Constructor Pook; and that the theories he has deduced from that study are correct ones, the fine ships built under his supervision at the government yards bear witness. The book will be of interest to every shipwright who looks higher than to the mere swinging of an axe."—*Chronicle, Portsmouth, N. H.*

ROGERS (H. D.) Geology of Pennsylvania. A complete Scientific Treatise on the Coal Formations. By Henry D. Rogers, Geologist. 3 vols. 4to, plates and maps. Boards 30.00

RUSSELL (J. Scott). The Modern System of Naval Architecture for Commerce and War. In Three Parts. Part 1, Naval Design. Part 2, Practical Ship-Building. Part 3, Steam Navigation. 1 vol. folio, 27 in. x 20 in., 724 pp. text, and 2 vols. plates, 165 in all, and engraved on copper, varying in size from folio double elephant, 27 in. x 20 in. to 27 in. x 90 in., and are drawn to a practical working scale. In Portfolio..............$60.00

In half Russia, 3 vols........................105.00

SHAFFNER (T. P.) Telegraph Manual. A complete History and Description of the Semaphoric, Electric, and Magnetic Telegraphs of Europe, Asia, and Africa, with 625 illustrations. By Tal. P. Shaffner, of Kentucky. New edition. 1 vol. 8vo, cloth, 850 pp.... 6.50

SILVERSMITH (Julius). A Practical Hand-Book for Miners, Metallurgists, and Assayers, comprising the most recent improvements in the disintegration, amalgama tion, smelting, and parting of the Precious Ores, with a Comprehensive Digest of the Mining Laws. Greatly augumented, revised, and corrected. By Julius Silversmith. Fourth edition. Profusely illustrated. 1 vol. 12mo, cloth 3.00

"'The Practical Hand-Book for Miners, Metallurgists, and Assayers,' by Mr. Julius Silversmith, a writer well known in connection with mining enterprises, forms a useful contribution to the scientific literature of the country. It will be found of equal value to those engaged in mining, either as actual workers, or as owners or shareholders in mining property. It gives a brief *resume* of the geology of metals, quite sufficient for practical purposes, and evidently gathered from the best authorities. It treats of mining in all its branches, for ores and native metals, classifying the several kinds of veins and deposits, and herein exhibits the modes in use in the exploration of mines, tunnels, shafts, adits, galleries, &c., and the timbering of mines This part of the book, as indeed most branches treated, is profusely illustrated by cuts, which must assist the ready comprehension of the matter discussed. A summary of the chemistry and metallurgical treatment of metals follows, and after giving descriptions, accompanied by engravings, of a large number of the new processes and inventions for treating ores, especially those of the precious metals, the book concludes with an exposition of the mining laws of the Pacific States and Territories."—*New York Herald.*

SILVER DISTRICTS OF NEVADA. 8vo, with map, paper 35

SMEDBERG. A Synopsis of British Gas Lighting, comprising the essence of the *London Journal of Gas Lighting*, from 1849 to 1868. Arranged and executed by James R. Smedberg, C. E., Engineer of the San Francisco Gas Works. Issued only to subscribers. 4to, cloth. (Ready soon.) $15.00

SPHERICAL ASTRONOMY. By F. Brunnow, Ph. Dr. Translated by the Author from the Second German edition. 1 vol. 8vo, cloth 6.50

STILLMAN (Paul). Steam Engine Indicator, and the Improved Manometer Steam and Vacuum Gauges—their Utility and Application. By Paul Stillman. New edition. 1 vol. 12mo, flexible cloth.............. 1.00

"The purpose of this useful volume is to bring to the notice of the numerous class of those interested in the building and the use of steam engines, the economy and safety attending the use of the instrument therein described. The Manometer has been long used—the inventor is Watt in a cruder form; and the forms herein described are patented by the author. The language of the author, the diagrams, and the scientific mode of treatment, recommend the book to the careful consideration of all who have engines in their care."—*Boston Post.*

SWEET (S. H.) Special Report on Coal; showing its Distribution, Classification, and cost delivered over different routes to various points in the State of New York, and the principal cities on the Atlantic Coast. By S. H. Sweet. With maps, 1 vol. 8vo, cloth.............. 3.00

STEWART (W. M.) The Mineral Resources of the Pacific States and Territories. By Hon. W. M. Stewart. 8vo, paper...................................... 25

WALKER (W. H.) Screw Propulsion. Notes on Screw Propulsion, its Rise and History. By Capt. W. H. Walker, U. S. Navy. 1 vol. 8vo, cloth........... 75

"After thoroughly demonstrating the efficiency of the screw, Mr. Walker proceeds to point out the various other points to be attended to in order to secure an efficient man-of-war, and eulogizes throughout the readiness of the British Admiralty to test every novelty calculated to give satisfactory results. * * * * Commander Walker's book contains an immense amount of concise practical data, and every item of information recorded fully proves that the various points bearing upon it have been well considered previously to expressing an opinion."—*London Mining Journal.*

WARD (J. H.) Steam for the Million. A popular Treatise on Steam and its Application to the useful Arts, especially to Navigation. By J. H. Ward, Commander U. S. Navy. New and revised edition. 1 vol. 8vo, cloth $1.00

WEISBACH (Julius). Principles of the Mechanics of Machinery and Engineering. By Dr. Julius Weisbach, of Freiburg. Translated from the last German edition. (In press.).................................

WHILDEN (J. K.) On the Strength of Materials used in Engineering Construction. By J. K. Whilden. 1 vol. 12mo, cloth 2.00

"We find in this work tables of the tensile strength of timber, metals, stones, wire, rope, hempen cable, strength of thin cylinders of cast iron; modulus of elasticity, strength of thick cylinders, as cannon, &c., effects of reheating, etc., resistance of timber, metals, and stone to crushing; experiments on brick-work; strength of pillars, collapse of tube; experiments on punching and shearing; the transverse strength of materials; beams of uniform strength; table of coefficients of timber, stone, and iron; relative strength of weight in cast-iron, transverse strength of alloys; experiments on wrought and cast-iron beams; lattice girders, trussed cast-iron girders; deflection of beams; torsional strength and torsional elasticity."—*American Artisan.*

WHITNEY (J. P.) Colorado, in the United States of America. Schedule of Ores contributed by sundry persons to the Paris Universal Exposition of 1867, with some Information about the Region and its Resources. By J. P. Whitney, of Boston, Commissioner from the Territory. Pamphlet 8vo, with maps. London, 1867. 25

——— Silver Mining Regions of Colorado, with some account of the different processes now being introduced for working the Gold Ores of that Territory. By J. P. Whitney. 12mo, paper........................ 25

WILLIAMSON (R. S.) On the Use of the Barometer on Surveys and Reconnaissances. Part I. Meteorology in its connection with Hypsometry. Part II. Barometric Hypsometry. By R. S. Williamson, Bvt. Lieut-Col. U. S. A., Major Corps of Engineers. With Illustrative Tables and Engravings. Paper No. 15, Professional Papers, Corps of Engineers. 1 vol. 4to, cloth. 15.00

ROEBLING (J. A.) Long and Short Span Railway Bridges. By John A. Roebling, C. E. Illustrated with large copperplate engravings of plans and views. *In press*...............................

CLARKE (T. C.) Description of the Iron Railway Bridge over the Mississippi River, at Quincy, Illinois. By Thomas Curtis Clarke, Chief Engineer. Illustrated with 27 lithographed plans. 1 vol. 8vo, cloth.................................... $7.50

TUNNER (P.) A Treatise on Roll-Turning for the manufacture of Iron. By Peter Tunner. Translated and adapted by John B. Pearse, of the Pennsylvania Steel Works, with numerous engravings and wood-cuts. *In press*....................

ISHERWOOD (B. F.) Engineering Precedents for Steam Machinery. Arranged in the most practical and useful manner for Engineers. By B. F. Isherwood, Civil Engineer, U. S. Navy. With illustrations. Two volumes in one. 8vo, cloth......... 2.50

BAUERMAN. Treatise on the Metallurgy of Iron, containing outlines of the History of Iron Manufacture, methods of Assay, and analysis of Iron Ores, processes of manufacture of Iron and Steel, etc., etc. By H. Bauerman. First American edition. Revised and enlarged, with an appendix on the Martin Process for making Steel, from the report of Abram S. Hewitt. Illustrated with numerous wood engravings. 12mo, cloth 2.50

"This is an important addition to the stock of technical works published in this country. It embodies the latest facts, discoveries, and processes connected with the manufacture of iron and steel, and should be in the hands of every person interested in the subject, as well as in all technical and scientific libraries."—*Scientific American.*

PEET. Manual of Inorganic Chemistry for Students. By the late Dudley Peet, M. D. Revised and enlarged by Isaac Lewis Peet, A. M. 18mo, cloth... 75

NUGENT. Treatise on Optics: or, Light and Sight, theoretically and practically treated; with the application to Fine Art and Industrial Pursuits. By E. Nugent. With one hundred and three illustrations. 12mo, cloth.................. .... $2.00

"This book is of a practical rather than a theoretical kind, and is designed to afford accurate and complete information to all interested in applications of the science.—*Round Table.*

BRÜNNOW. Spherical Astronomy. By F. Brünnow, Ph. Dr. Translated by the Author from the Second German edition. 1 vol., 8vo, cloth............. 6.50

GLYNN (J.) Treatise on the Power of Water, as applied to drive Flour Mills, and to give motion to Turbines and other Hydrostatic Engines. By Joseph Glynn. Third edition, revised and enlarged, with numerous illustrations. 12mo, cloth............ 1.25

PRIME. TREATISE ON ORE DEPOSITS. By Bernhard Von Cotta. Translated from the Second German edition by Frederick Prime, Jr., Mining Engineer, and revised by the Author. With numerous illustrations. *In press*........................

HUMBER. A Handy Book for the Calculation of Strains in Girders and similar Structures, and their Strength, consisting of Formulæ and corresponding Diagrams, with numerous details for practical application. By William Humber. 12mo, fully illustrated, cloth... 2.50

GILLMORE. Engineer and Artillery Operations against Charleston, 1863. By Major-General Q. A. Gillmore. With 76 lithographic plates. 8vo, cloth ... 10.00

——— Supplementary Report to the above, with 7 lithographed maps and views. 8vo, cloth............ 5.00

MILITARY AND POLITICAL LIFE OF THE EMPEROR NAPOLEON. By Baron de Jomini, General-in-Chief and Aid-de-Camp to the Emperor of Russia. Translated from the French, with Notes by H. W. Halleck, LL. D., Major-General U. S. Army. 4 vols. royal 8vo, with an Atlas of 60 Maps and Plans, cloth $25.00

Half calf or morocco 35.00

Half russia 37.50

TREATISE ON GRAND MILITARY OPERATIONS. Illustrated by a Critical and Military History of the Wars of Frederick the Great. With a Summary of the most important principles of the Art of War. By Baron de Jomini. Illustrated by Maps and Plans. Translated from the French By Col. S. B. Holabird, A. D. C., U. S. Army. In 2 vols. 8vo, and Atlas, cloth 15.00

Half calf or half morocco 21.00

Half russia 22.50

CAVALRY: ITS HISTORY, MANAGEMENT, AND USES IN WAR. By J. Roemer, LL. D., late an Officer of Cavalry in the Service of the Netherlands. Elegantly illustrated, with 127 fine wood engravings. In one large octavo volume, beautifully printed on tinted paper. Cloth 6.00

Half calf 7.50

ELEMENTS OF MILITARY ART AND HISTORY. By Edward de la Barre Duparcq, Chef de Bataillon of Engineers in the Army of France, and Professor of the Military Art in the Imperial School of St. Cyr. Translated by Brigadier-General George W. Cullum, U. S. A., Chief of the Staff of Major-General H. W. Halleck, General-in-Chief U. S. Army. 1 vol. 8vo, cloth 5.00

www.ingramcontent.com/pod-product-compliance
Lightning Source LLC
LaVergne TN
LVHW011213110826
845150LV00006B/1423

* 9 7 8 1 4 2 5 5 1 7 2 4 3 *